Conscious Mind in the Physical World

Conscious Mind in the Physical World

Euan Squires

Department of Mathematical Sciences
University of Durham

CRC Press
Taylor & Francis Group
Boca Raton London New York

CRC Press is an imprint of the
Taylor & Francis Group, an **informa** business

CRC Press
Taylor & Francis Group
6000 Broken Sound Parkway NW, Suite 300
Boca Raton, FL 33487-2742

First issued in paperback 2019

© 1990 by Taylor & Francis Group, LLC
CRC Press is an imprint of Taylor & Francis Group, an Informa business

No claim to original U.S. Government works

ISBN-13: 978-0-7503-0045-2 (hbk)
ISBN-13: 978-0-367-40327-0 (pbk)

British Library Cataloguing in Publication Data

Squires, Euan
    Conscious mind in the physical world.
    1. Consciousness
    I. Title
    126

Library of Congress Cataloging-in-Publication Data are available

**Visit the Taylor & Francis Web site at
http://www.taylorandfrancis.com**

**and the CRC Press Web site at
http://www.crcpress.com**

René Descartes, who was the first "modern" thinker really to appreciate the problem which lies at the heart of this book, and whose influence, for better or worse, has pervaded the subject for over three hundred years, once wrote that a beautiful woman, a good book and a perfect preacher were the things most impossible to discover in the world. I doubt very much whether he would regard this as a good book, although he would certainly be amazed by what he discovered in it, and would, I believe, approve of its aims. I have not been able to resist the temptation to include what he would surely consider to be a very imperfect sermon, and I dedicate the book to Eileen.

Contents

x   *Contents*

Acknowledgements

I am grateful to Adam Hilger referees for comments and criticisms of an earlier draft of this book; I hope these have resulted in many improvements. I would also like to thank several other people with whom I have, at various times, discussed some of the problems of quantum theory, in particular, John Bell, Yoav Ben-Dov, Peter Collins, Lucien Hardy, Ian Lawrie, Nick Maxwell and Tony Sudbery.

Chapter 1

Introduction

1.1 The book

This book is concerned with what is surely the most obvious and yet the most mysterious, and hence the most fascinating, of all phenomena: the conscious mind. The emphasis here is on the word "conscious"; we are not primarily concerned with the mechanical working of the brain, i.e. how it receives, stores and processes information, or with how such working can be modelled on a computer, interesting though such topics are, and in spite of their more obvious relation to physics. Our concern is with consciousness itself; not, for example, with the way an image is formed on my retina and then transmitted to my brain, but with the fact that I am aware of the sensation of seeing.

An attempt will be made to discuss consciousness within the context, and using the methods, of physics, or, more precisely, theoretical physics. Indeed much of the book is simply an account of some features of modern physics. Given the background of the author, as explained in the next section, this is perhaps inevitable, but it raises two immediate questions. One is whether conscious mind is within the domain of physics, i.e. should physicists, as physicists, be interested in the subject, and the other is whether physics has anything useful to contribute. Attempts to provide at least partial answers to these questions will be found in the following chapter.

Although the issues with which this book is concerned are extremely deep, profound and difficult, the basic questions can be presented, at least initially, in reasonably simple terms. The difficulties come when we try to make the questions more precise; still more when we try to answer them! I have therefore tried to keep the discussion at an elementary and introductory level. There are a few mathematical equations in the text, but I hope readers who are unhappy with such things will be able to ignore them, and still follow the argument. In the more "philosophical" parts I have had no difficulty in avoiding technicalities (I would be incapable of understanding them). Some readers may feel that I have oversimplified the issues. Undoubtedly the work of many people, published in thousands of pages, has been reduced to a few words, and almost every chapter could have been expanded to make a complete book. In my defence I can refer to a desire to keep the discussion brief, to a general feeling that some philosophical discussions have a length that is disproportionate to their content and (since the last remark is too provocative) to my lack of experience in the subject, which means that it would be presumptuous of me to attempt anything other than an introduction.

The chapters in this book are intended to follow each other in a logical sequence. I have tried, however, to make each chapter reasonably self-contained, even at the cost of some repetition. I hope this will make the book easier to read, and will allow a certain amount of reader-selection in the order of chapters.

Very little in this book is claimed to be original. Essentially all of it has surely, in some form or another, been written elsewhere. Certainly the basic questions have been discussed, in various guises, since thinking man evolved. Of course the context has changed, and one of the purposes of this book is to study the questions in the light of our modern understanding of the physical world. It is a fact, still not properly appreciated by many scientists and philosophers, that quantum phenomena have revolutionised our view of that world. This revolution should not be ignored in any serious discussion of conscious mind. I have included some details of recent developments in the continuing endeavour of trying to interpret quantum theory, particularly as these are not widely known. In particular, I have tried to show how one model, closely related to the so-called *many-worlds* interpretation, might suggest answers to at least some of the problems regarding consciousness. These answers could have significance for topics that go far beyond the "merely philosophical". The last chapter is an attempt to provide a few conclusions, and is entirely a personal view.

For whom is this book written?

This is the question publishers always ask (rightly so) and authors tend to ignore. (It is easier to write the book we want to write, rather than the one somebody might want to read!) There are three principle groups to whom I hope it might be of interest. Firstly, there are psychologists, neuroscientists and, in particular, philosophers, who, in different ways, have a professional interest in the study of the mind and who, I believe, should know something about physics. A quick survey of the many books on the topic of the mind in the library here in Durham shows that this view is not generally held. It is clear that most authors consider physics to be irrelevant. Quantum theory is not mentioned until p.537 of the excellent and detailed study given by Gregory (1981) in *Mind in Science*, and then only in a very cursory way; it is essentially ignored in *The "Mental" and the "Physical"* (Feigl, 1967) and in *The Philosophy of Mind* (Smith and Jones, 1986); and it does not even appear in the index of *Matter and Consciousness* (Churchland, 1984), where I read: *The phenomena to be penetrated are now the common focus of a variety of related fields. Philosophy has been joined by psychology, artificial intelligence, neuroscience, ethology, and evolutionary theory, to name the principals.* There is no mention of physics. These, and many others, are excellent books, and readers will find in them much more detailed discussion of the issues raised in chapter 5, for example, but their lack of interest in physics is strange. How is it possible to argue about materialism, without some words about the nature of "matter", or about physicalism, unless there is some agreement about what "physics" actually is? As we shall see, these things are not trivial.

Much of this book is concerned to explain those aspects of physics, in particular quantum physics, which might be relevant to the mind. Most physics text-books are, quite properly, concerned with other aspects and applications of their subject, and are not ideal for this purpose. In consequence, physics is sometimes considered to be too daunting, or, what is worse, knowledge is acquired from unreliable sources. At the risk of again being provocative, I can quote as examples the reluctance of Popper and others to appreciate the implications of interference in quantum theory; the obsession that is sometimes seen with regard to the uncertainty principle, which is more properly regarded as a trivial consequence of the theory than as a fundamental principle; the failure to recognise that the most significant revolution brought about by quantum theory is *not* the breakdown of causality; and, lastly, what seems to me the total misunderstanding of the subject found in books like *How the Laws of Physics Lie* (Cartwright, 1983).

Secondly, I hope that physicists will be interested. They will not learn anything new about how to calculate things in physics, but many will find the discussion of quantum theory goes beyond what they have learned from text-books and lectures, and I hope they will be fascinated by a topic that ultimately is too important to be left to philosophers. Their problems are often different to ours.

Finally, there is a large group who would not claim to be experts in any of the above subjects, but who are interested in what is known of the conscious mind, and, in particular, in whether it can be regarded as part of the physical world, or whether it inevitably requires something that is beyond physics. These are the issues which we shall continually be meeting.

1.2   The author

Since our only direct experience of a conscious mind is of our own consciousness, there is perhaps more excuse here than in most books for the author to say a little about himself.

Many years ago at the University of Manchester, I trained as a theoretical physicist. The head of the department was Leon Rosenfeld, who was extremely interested in the philosophy of science and who even lectured on it to his students. Unfortunately, I understood very little, and so, apart from the fact that he was an admirer of the Greek philosopher Epicurus, born in Samos in 341 BC, the only thing I remember is that he gave us what I now believe to be a mistaken idea that the problems of quantum theory had been solved by Niels Bohr. Although I am sure none of us comprehended the solution, we were afraid to admit it!

Since 1964 I have been professor of applied mathematics at the University of Durham. Here, apart from teaching students, I have endeavoured to do research into the theory of elementary particles. In practice this has largely meant endeavouring to keep up with the progress being made by other research workers, and, without any doubt, I have been fortunate to live through a period of time when such progress has been both rapid and exciting.

To the obvious protest that the above background gives me no qualification to write about the subject of conscious mind, my immediate reaction is to say that so little is known about the subject that I am as well qualified as anyone. Of course the topic has been central to philosophy for at least two centuries, so my claim that little prior knowledge is required might be disputed. That I see little content in the work of so many clever people does not necessarily mean that more is not there. However, philosophers of science have also written extensively on a subject about which I do know something, namely, quantum theory, and it is clear that on this topic a lot of words has produced few things of significance. When philosophy has moved into physics, it has not made much progress—another provocative remark, and one which clearly invites a similar response to this book, in which a physicist is trying to move into philosophy!

I must also qualify the above statement that little is known on the subject of conscious mind, by mentioning the remarkable recent work of neuroscientists, some of which is discussed briefly in chapter 6. The problem here is to know whether the results really have any relevance to our topic. In any case, I am encouraged by the fact that John Eccles, who was awarded a Nobel Prize for neuroscience, recently wrote: *Unfortunately it is rare for a quantum physicist to risk an intrusion into the brain–mind problem* (Eccles, 1987, p.301). I take that to be an invitation, which I am glad to accept.

The third group of people who have professional interests related to conscious mind are psychologists. However, partly in a desire to be regarded as *serious* scientists, many psychologists try to avoid using the language of consciousness, and instead concentrate strictly on observation and description of behaviour (see section 5.4).

The idea of writing this book grew out of a previous semi-popular book on quantum theory (Squires, 1986), in which I was inevitably forced to write about consciousness, for reasons that will become clear in chapter 10. This led me to read about the subject, and the more I read, the more I became both fascinated and dissatisfied. Part of the reason for the latter feeling is presumably that what I read was not written by theoretical physicists, who have provided most of my reading material for the past 30 years, and who have a way of writing and thinking about problems which is different to that of the world in general, perhaps even to that of most scientists, but which, given the success they have had, is surely of value. Of course many would argue that such a way of thinking is not appropriate to the topic of this book and, to some extent at least, they would be right.

In writing about conscious mind I am also aware that I am trying to do a little bridge-building. The cultural gulf between the scientific way of thinking, seen most clearly in the discipline of theoretical physics, and the way of thinking of all non-scientists, and indeed generally of scientists when they are not doing science, is vast. The basic assumptions, the accepted "paradigms" (an ugly word which I promise not to use again), are incompatible, or, perhaps it is truer to say, are so different that one set is meaningless to the other. I hope that at least some glimpses of "the other side" will be seen.

A final motivation is the desire we all have to answer the question: *Who am I?* I do not of course know the answer; if I did this would be a very different book. The most I can expect is that I might have clarified some of the questions. This last sentence may indicate another difference between the way physicists and philosophers write. In general I am not trying to propagate a particular position, or "ism". Indeed, as we shall see, I suspect that some of the strongly held views are not in fact so different to each other as their advocates apparently believe.

Just over 40 years ago one of the founders of quantum theory, E Schrödinger, prefaced a book entitled *What is Life?* (Schrödinger, 1944) with the words:

> ... *we are only now beginning to acquire reliable material for welding together the sum total of all that is known into a whole; but, on the other hand, it has become next to impossible for a single mind fully to command more than a small specialised part of it. I can see no escape from this dilemma than that some of us should venture to embark on a synthesis of facts and theories, albeit with secondhand and incomplete knowledge of some of them—and at the risk of making fools of ourselves.*

The present book is written in a similar spirit.

Chapter 2

Physics and conscious mind

2.1 Physics as the theory of everything

In this section we ask whether conscious mind can properly be regarded as being in the domain of physics. Clearly we would answer such a question in the negative if we defined physics to be that which deals with the "physical", and then adopted the view that conscious mind is "non-physical". This answer would seem to require accepting already a so-called dualist position (see section 5.6), i.e. we would be separating the real world into two parts and designating one as belonging to physics.

Here, instead, and without at this stage wishing to prejudice the question of whether some sort of dualism is correct, we shall avoid making such an arbitrary restriction to what we call physics and take the positive view based on the fact that physics is generally regarded by those who study it as being (uniquely) **the** fundamental science. The prime task of fundamental physics is to understand the objects, laws, or whatever, that are the basis of *all* observed phenomena.

Later we shall discuss, and, I hope, be encouraged by, the enormous success that fundamental physics has had in such an endeavour. It is very impressive. The progress that has been made in this century is such that physicists sometimes dare to speak of a TOE, a "theory of everything". An honest assessment would say that such a TOE is, in reality, very far off (and in any case I am not sure that I am completely clear what the idea means), but at least it can be discussed. A characteristic of this is that, within a group of contemporary physicists, there are no questions that are considered unreasonable; we can ask why space has three dimensions, why time has only one, why the charge on an electron has the value it has, why there are so many types of quark, why the galaxies are in clusters, etc. We may not be able to *answer* these questions, and it is probable that there are better questions that we have not even thought of, but at least there is the feeling that answers are possible.

However, even if we were able to find a theory which explained all observed phenomena, it would say nothing about the basic process of observation, i.e. the process through which there are any phenomena to be explained. *Everything* we know, we know by means of the conscious mind. A theory of everything would certainly explain why the light emitted from a particular transition of the sodium atom had a wavelength 5.89×10^{-7} m, but I perceive this light as "yellow", and present physics does not contain such a concept.

What do we deduce from this? There have been many other times in the history of physics when the subject has suddenly been confronted with new phenomena. Electromagnetic waves, discussed in section 4.1, are a good example. However, in general, and usually in a short time, the new discoveries have been incorporated into an expanded version of physics. Should we expect that the same thing will happen with regard to consciousness? Will the properties of the conscious mind one day be included among the things which are discussed in physics text-books? Maybe the most reasonable guess is to answer no. But why?

It is important now to emphasise that we should not be seduced into accepting what, at least to the scientific community, might seem to be an easy way out. This runs roughly as follows (though it would never be explicitly stated in this form): physics is wonderfully successful in explaining the properties of the real world; physics does not contain anything about consciousness; hence consciousness is somehow less "real" than the things of physics. There is no logical basis for such an argument and it involves a rather arrogant view of our scientific attainments— *"what I don't understand isn't real"* is no more reasonable than *"what I don't know is not knowledge"* (attributed, as a joke, to the Master of Balliol College by H C Beeching)—but nevertheless it is necessary to be explicit in rejecting it.

As a theoretical physicist I believe, for example, in the existence of quarks. The evidence on which this belief is based is extremely indirect; it comes to me personally only through having read articles written by people who have done the relevant experiments or who have provided the detailed interpretation of these experiments, perhaps even mainly through having heard talks by people who themselves have read these articles. Although I might not myself be able to repeat many of the crucial steps in the argument, I accept that they have been done properly. The information is essentially "n^{th}-hand", where n is quite large. The evidence for conscious experience, however, is totally personal; I do not have to trust the skill, or the reliability, or the integrity of any person to *know* the reality of redness, or fear, or love, or happiness, etc. It is impossible to define what we mean by "being real" or "existing", but on any reasonable interpretation of these words, that which I experience cannot be less real, or have less claim to exist, than that which I deduce. Nobody who wishes to be interested in a theory which tries to explain everything, can be wholly unconcerned with the conscious mind.

Of course it is possible that we will never be able to regard consciousness as a part of physics. Maybe it is so qualitatively different from the things that we deal with in physics, that it will be forever separate. Such a conclusion would itself be a very interesting result, and, as we noted earlier, would invite the question: Why? What could there be in the definition of what we want to call physics which would make the inclusion of consciousness impossible? There is no escaping such fascinating questions. Pippard (1988) has suggested that the crucial distinction between physics and the things of the conscious mind is that whereas the former belong to the "public" domain, and can therefore be discussed, the latter belong entirely to the "private" domain, which is closed to physics. This is a very interesting distinction, but of course if we did have a physics of consciousness, then it too would enter the public domain, so it is not clear to me that we are not here assuming the answer.

I will add two further comments relevant to the question of whether physicists should be interested in conscious mind. The first is that one of the greatest physicists of the 19th century, James Clerk Maxwell, once expressed the opinion that atoms would be forever outside physics; they were the God-given objects with which physics dealt, but neither they nor their properties would ever be understood. As we shall see in chapter 4, he was wrong (not just a little bit wrong, he was *totally* wrong). We should not repeat his mistake in a different context. Nobody in 1889 could have guessed that the physics of the 20th century would be the physics of atoms and their constituents. It is unlikely that in 1989 we are any better equipped to guess at the topics of the physics of the 21st century. It just *might* have something to do with the conscious mind.

The other comment is personal. I know at least one physicist who is interested in consciousness, namely, myself. Some of my friends probably believe that I am writing about it because TOE's have become too difficult. They may be right.

2.2   Physics and the violation of experience

In the last section we looked at reasons why physicists, as physicists, should be interested in the conscious mind. Now we consider the remainder of the human population, i.e. the vast majority who are not physicists. In general they will want to set up a barrier; to say "keep-out" to those who seek to bring physics into conscious mind. Opposition to the claims and methods of physics occurs at many levels. It is seen in the writings of anti-realist philosophers, who would deny to physical theories any claims to be true or false or to provide explanations of observed phenomena (see section 5.2 for further discussion of such ideas). At the more popular level it is evident from the success of any book that tries to demolish the claims of science, more particularly of physics, in favour of some mystical, pseudo-scientific, cult (ranging from the reasonably cogent to the absurd); in the correspondence columns of newspapers; in the sad growth of astrology, about which the 1875 edition of the Encyclopedia Britannica wrote that it was *"no longer necessary to protest against an error which is dead and buried"*, and related branches of the fortune-telling industry; in the fact that in the most technologically advanced society in the world more people believe in UFO's than in Darwin's theory of evolution; and in the general feeling that we would all rather *like* to believe that Uri Geller can really bend spoons by thinking suitable thoughts, and

so make physicists appear foolish. In its extreme version this opposition sees modern science, which is founded on the methodology
of physics, as the root of all evil: *It ... expounds the thesis that the
scientific mentality is the source of the evils of the modern world.*
(These words appear in P E Hodgson's review (Hodgson, 1988) of
The Rape of Man and Nature by P Sherard.)

There are several reasons for this general anti-physics attitude,
an attitude which I believe to be potentially harmful for individuals and society. Undoubtedly jealousy plays a part in some cases.
Physicists have been very successful, not only in the advance of
their subject, but also in obtaining money from governments to
continue it! They have been, perhaps, a little arrogant in regarding everything else as being less fundamental, and therefore less
interesting (?). There may also be the feeling that modern theories in physics are so difficult and impenetrable to outsiders (they
are—see remarks at the close of the previous section) that it would
be very pleasant if they were all shown to be completely wrong, so
that the many clever people who claim to understand them were
made to seem more human.

Another reason behind anti-physics attitudes originates from a
concern for the environment. Such a concern is certainly justified and indeed we need more of it, but I believe that it is wrong
to attribute blame for damage to the environment to physics, or
more generally to science. Although it is true that scientific understanding does allow us to misuse our environment, it is ultimately
only through such understanding that we will be able to protect
and preserve it. A similar response can be made with respect
to the military uses of science. Whilst science and its resulting
technology allow us to construct ever more powerful weapons of
destruction, they also make such weapons increasingly irrelevant
because they enable us to see that we live in one world, a world
which properly utilised can provide for the wants, and more, of all
its people. These facts of course are not yet properly appreciated
by politicians, but sometimes I think that, even there, there are
signs of hope. In the words of the "Erice Statement" (1982), signed
by over ten thousand international scientists: *The choice between
peace and war is not a scientific choice. It is a cultural one: the
culture of love produces peaceful technology. The culture of hatred
produces instruments of war.*

One other source of opposition to physics, in particular to

physicists being interested in the conscious mind, is more directly
relevant to our topic. This can perhaps best be regarded as a type
of "fear". It is the fear expressed when the analytical mind of
the philosopher Apollonius intrudes into young love's *purple-lined
palace of sweet sin* in John Keats' poem *"Lamia"*:

> ... *Do not all charms fly*
> *At the mere touch of cold philosophy?*
> *There was an awful rainbow once in heaven:*
> *We know her woof, her texture; she is given*
> *In the dull catalogue of common things.*
> *Philosophy will clip an angels wings,*
> *Conquer all mysteries by rule and line,*
> *Empty the haunted air, and gnomèd mine-*
> *Unweave a rainbow, as it erewhile made*
> *The tender-personed Lamia melt into a shade.*

Of course by "philosophy", he really meant "physics"! Put simply,
the fear is due to the fact that physics seeks to explain *everything*
in terms of the motion of elementary particles moving inexorably
according to the rules of mechanics—**there is nothing else.** In
such a world what place is there for the things that mean the most
to us; for love and beauty, for truth and honour, for freedom and
responsibility, for joy and hope, ..., or indeed for the converse
of all these things? They somehow seem to become empty and
meaningless illusions. Steven Weinberg, who played a vital role in
discovering the extremely successful "standard model" of contemporary physics (section 4.4), finds no escape from this depressing
conclusion. Towards the end of a book on the early universe, *The
First Three Minutes* (Weinberg, 1977), he writes: *The more the
universe seems comprehensible, the more it also seems pointless.*
Such words, however, already contain their own apparent contradiction; the fact that somebody *cares* whether the world has meaning is surely sufficient to imply that it has. In a slightly different
context Coleridge wrote about the same problem:

> *If rootless thus, thus substanceless thy state,*
> *Go, weigh thy dreams, and be thy hopes, thy fears,*
> *The counter weights!—Thy laughter and thy tears*
> *Mean but themselves, each fittest to create,*
> *And to repay the other! Why rejoices*
> *Thy heart with hollow joy for hollow good?*

Why cowl thy face beneath the mourner's hood,
Why waste thy sighs, and thy lamenting voices,
Image of Image, Ghost of Ghostly Elf,
That such a thing as thou feel'st warm or cold?
Yet what and whence thy gain, if thou withhold
These costless shadows of thy shadowy self?
Be sad! be glad! be neither! seek, or shun!
Thou hast no reason why! Thou canst have none;
Thy being's being is contradiction.

Physics seems to threaten the very substance of what "I" am, it seems to deny the "me-ness" of being me. We have to find a way to defend ourselves from this brazen attack on the citadel of our conscious experience. If we are too honest to do this by refusing to listen to the claims of physics, we invent "level-autonomy" (see section 2.3) to prevent physics from going beyond the limits of the microscopic, or indeed we might follow Whitehead who created a complete philosophical theory (see section 5.8) to protect us from *the violence done to experience by the idea that physics is metaphysically primary* (W E Hocking in *The Philosophy of Alfred North Whitehead* (Schlipp, 1941)). Part of the reason for this book is to try to face some of these issues, and not just run away from them. I think that shouting slogans like "level-autonomy" *is* running away. The same may be true of Whiteheadean philosophy, but I do not understand it sufficiently well to make a judgement. We shall not need to run away because we shall remain confident that the citadel is secure; as we have already confirmed, the things of conscious mind *are* real. If a suitable version of physics contains and explains them then that is part of the wonder of what physics is. If, on the other hand, it cannot, then there is something else, something beyond the world of physics.

2.3   Physics and reductionism

An important ingredient of the methodology of physics, which has been crucial to almost all the progress made this century, is that of taking things apart in order to discover of what they are made. In this way we have seen how the properties of large complex objects are a consequence of the (simpler) properties of their constituents. Probably the best example of this from physics is the success, noted earlier as exceeding Maxwell's expectations, in explaining the properties of atoms from those of their constituent electrons. The everyday world of course provides many other obvious examples. Even the micro-computer, on which I am writing these words, works as it does because of the properties of the many parts of which it is constructed.

A physicist might therefore approach the subject of conscious mind by asking the questions: Of what is it made? What are the key constituents which are required before an object becomes conscious? Again these may be silly questions, but if they are it is important to know why.

The name given to the process of trying to understand objects through the properties of their constituents is **reductionism**. This word arouses very strong feelings and has, unfortunately, become for many a term of abuse. A recent book on quantum theory makes the point very well: ... *in principle one can envisage a chain of analysis in which sociology is analysed into psychology, psychology into physiology, physiology into biology, biology into chemistry and chemistry into physics. (This view of science is called 'reductionism' by those who don't like it and 'the unity of science' by those who do)* (Sudbery, 1986). To call an argument "reductionist" is sometimes used, without any valid reason, as a way of rejecting it! Here we meet again the fear noted in the previous section; to "reduce" everything to physics is seen as losing all the things we cherish most. Partially as a consequence of this fear, reductionism has been attacked from diverse directions, often with considerable passion, and with arguments that have an emotive content greatly exceeding their rationality.

In the early 17th century the intrinsically reductionist notion of "atomism", i.e. that material substances are made from specific atoms (the distinction between atoms and molecules was not appreciated at this time), was opposed by some factions of the catholic church because it was considered to be incompatible with a jealously guarded dogma. At the Last Supper of Jesus of Nazareth with his friends, immediately prior to his crucifixion, he is reported to have broken a piece of bread whilst saying: *"this is my body"*. Although it would seem reasonable to regard such words as being figurative, this has not been the traditional view of most branches of the Christian church, which have rather taken them extremely literally, and even extended them to imply that at each commemoration of this supper in the Holy Communion sacrament, the bread used actually *becomes* the physical body of Jesus. The precise nature of this supposed "transubstantiation" has been the subject of endless debates; indeed wars have been fought over the issue, and people have been burned at the stake for having allegedly defective understandings of the doctrine (such is the folly of those who would regard themselves as wise). For reasons that are again somewhat hard to understand, the church authorities believed that the doctrine was more easily acceptable within an Aristotelian version of reality, where "substance" had a more open meaning, than within atomism, where there lingered the uncomfortable fact that if the bread contained atoms of flour, etc, then it really *was* bread! It is interesting to note that much of the opposition to Galileo was probably due more to his belief in atomism than to his support for Copernicus. After all, the relative motions of the sun and the earth are not so deeply involved in church doctrine as the sacraments (see Redondi, 1988).

From a totally different (Marxist) viewpoint, and from a different age and culture (our own), I read ... *reductionism resorts to more or less vulgar simplifications which, in the prevailing social climate, become refracted into defences of the status quo in the form of biological determinism, which claims that the present social order, with all its inequalities in status, wealth and power between individuals, classes, genders and races, is 'given' inevitably by our genes. This limit to the scientific vision is compounded by the closed recruitment process into science as an institution which effectively ensures its preservation as the privilege of the western white male* (Rose, 1987).

In this book we shall regard reductionism as a *method*, not as a *belief* (scientists, as scientists, are not really supposed to be influenced by beliefs but by evidence). The method always depends upon trying to explain the properties of an object in terms of the known properties of its constituents, which are in some sense smaller and simpler. If this proves impossible then two reasons immediately suggest themselves:

either the composite object contains additional constituents that we failed to include;

or we did not correctly understand the properties of the constituents.

Are there other possibilities? It is certainly hard to imagine what these might be, although, as we shall see, there are subtleties within quantum physics that make the issues somewhat more obscure than at first sight they might appear. Nevertheless, it is not satisfactory to dismiss a reductionist argument without having some idea *why* it is wrong. It is easy to say that we cannot account for human emotions, for example, by a model of a man made entirely out of physical things, and all our instincts would encourage us to agree with the statement. The physicist within me, however, is not satisfied with this unless I know why.

There is an open-endedness about reductionism; it asks questions. Maybe this is the reason why it is disliked by dogmatists of all persuasions.

It is necessary now that we should try to understand more carefully what we mean by saying that the properties of a complex object are a consequence of the properties of its constituents. In particular, we do not mean that "organisation", i.e. the way things are put together, is not important. At a basic level, a computer, a car, a book and a sugar crystal are all made of identically the same constituents; they are just put together differently. As a more trivial illustration, there are many children's building toys, e.g. Lego, from whose pieces a great variety of different objects can be constructed. There is, however, nothing in their properties that is not a direct consequence of the particular arrangement of their constituents. I can assert, with reasonable confidence, that the same thing is true of a computer and the other things mentioned above; they are a consequence of the properties of quarks and leptons (see section 4.4). In other words, if I was clever enough (and had a large enough computer to help me), I could calculate how these things would behave, to arbitrary accuracy, from knowledge of experiments done on the constituents.

The only (minor) doubt I have in the above statement arises from the strange properties of quantum theory. As we shall see, it is not completely obvious what the statement means. Nevertheless, the success that the general reductionist approach has had surely means that those who wish to reject it should produce proper reasons, and not mere prejudice.

That the qualification "if I was clever enough" is necessary in the above discussion does not seriously weaken the reductionist statement. It would indeed be very arrogant for me, or anyone else, to claim that there was any *fundamental* difference between the things I am actually able to calculate and the things that are just too difficult. The only fundamental difference is between things that are in principle calculable and things that are not. Thus, as we shall see, the spectral lines of all atoms can be calculated from an equation which is very simple to write down, but which becomes rapidly far too complicated to solve, for all but the lightest atoms, except by using very crude approximations. On the other hand, the calculation of "radiative corrections" to the values obtained in the above way is *in principle* impossible, unless I include additional effects arising from relativistic field theory.

Another important point to notice is that a composite object may have properties that are not in themselves properties of the constituents, and may indeed not have any meaning for the constituents, without in any way invalidating the reductionist method. As an example, when we put together molecules of a particular type the result may be a solid or a liquid or a gas. These properties, of course, have no meaning for the individual molecules; they are "emergent" properties, which are only relevant when studying a large collection of molecules. Nevertheless, whether the resulting material is solid, liquid or gas is in principle calculable from the properties of the molecules; it is not in this sense something extra that has to be added in order to understand the substance.

There appears to be much confusion in some of the things that are said on this topic. For example, Popper writes ... *new atomic arrangements may lead to physical and chemical properties which are not describable from a statement describing the arrangement of the atoms, combined with a statement of atomic theory* (Popper and Eccles, 1977, p.23). This assertion is surely false. If we found that a substance containing a particular arrangement of atoms had a property that was not, in principle, deducible, then, first, we would be astonished and, secondly, we would conclude that in fact the substance contained *something else*. (There are indeed many cases in the history of science where this sort of thing has occurred. Note that the important thing is again whether the property is deducible, not whether it had in fact been deduced. Experiments have on occasions surprised us with effects that we could have predicted, but had failed to do so.) Later in the same article Popper tries to find examples to support his assertion and suggests that the lifetimes for the decay of various nuclear states are likely candidates. However, whilst these are often difficult to calculate in detail, there is no reason to believe that they do not follow directly from the structure of the relevant nucleus. In particular, it is easy to find a qualitative understanding of the enormous range of observed lifetimes, contrary to Popper's claim that such a range provides evidence against the reductionist method.

We have already mentioned the idea of *level-autonomy*. This is the claim that each level of existence, e.g. the subnuclear, the nuclear, the atomic, the inorganic, the organic, etc, is only partially explicable by the lower levels, and that "something new" is introduced at each level. If by this it is implied that there is some new extra ingredient, then there is nothing here with which a convinced reductionist would not agree. We cannot expect an object D to be explained completely in terms of the properties of its constituents A and B, if D also contains a constituent C. However, it is not clear whether those who use the expression "level-autonomy" really do mean this; certainly there is a reluctance to say what the new things are. Sometimes it seems as if they use it more as a slogan than an argument, to discourage reductionist explanations for no other reason than that they do not like them!

There is an important distinction that we must now make; and one that is often ignored. The statement that an object can be understood in terms of its constituents is not the same as saying that we understand why the object should in fact *exist*. Thus to assert, as I do with some confidence, that the working of a bicycle can be understood in terms of the component parts, wheels, chains, handle-bars, etc, of which it is constructed, does not mean that I have any understanding of why there should be any bicycles in the actual world.

Having looked at a few of the qualifications, we can return to the basic idea of the reductionist method, and repeat our claim that it will work for sodium chloride, for a sugar crystal, for a bicycle, for an aeroplane, etc. With almost equal confidence we can expect it to be valid for a plant, for a piece of a human body, e.g. a heart or a kidney; but for an animal, or a man, our doubts begin to dominate and we are left only with questions. Is a *person* completely explainable in terms of the particles of which he is made or is there *something else*? More particularly, is consciousness a property of some special arrangements of quarks and leptons (and if so what are the key features of these arrangements), or is it not? It seems as though most people who have thought about these questions fall readily into one of two groups, those who answer definitely yes, and those who with equal confidence answer no, having in common only the view that the opposite opinion is nonsense and hardly worthy of consideration!

The fact that in the above discussion we jumped from obvious "machines" to the conscious mind means that we have ignored the claims of, for example, a piece of music, or a sunset, to possess something, beauty, which is not a consequence of their constituents. I think this is a reasonable thing to have done because, in fact, beauty, etc, are not intrinsic properties of the objects themselves, but depend for their existence on conscious mind. When Paul Davies writes: *(Such a claim) is as ridiculous as asserting that a Beethoven symphony is nothing but a collection of notes* (Davies, 1983, p.62) he presumably is not meaning to say that the sounds made during the playing of the symphony cannot be exactly analysed into a set of particular frequencies at particular intensities, leaving nothing else. In so far as there is anything else to the symphony it is because of how it is appreciated by the conscious mind of somebody who hears it. Without a conscious mind a symphony really is *nothing more* than a collection of notes. (Even this remark needs clarification; one of the reviewers who read an early version of this book said it was too strong, another that it was too weak! The music can be exactly analysed into sound waves, which can be described mathematically, and indeed the whole performance can be reduced to digital form, as is done in certain types of recording. The mathematical description will contain certain recognisable structures, but these are not anything *extra* to the notes. It is only because of conscious mind that the particular sequences acquire the significance of being music.)

Finally, we mention two examples in physics where simple reductionism does in fact break down. The first concerns the *exclusion principle* of quantum theory. This asserts, for example, that we cannot have more than one electron in the same quantum state, a fact which we clearly could not predict from experiments on one electron, but which we might begin to guess from experiments on two or more. We could also deduce the result if we knew the quantum field theory that describes these electrons. The other example is the second law of thermodynamics which, as we shall see in chapter 8, does not follow from the laws of physics and which indeed is somewhat surprising, but which is probably a consequence of the "initial conditions" at the start of the universe, or which might be due to our living at a particular period in time.

2.4   From physics to conscious mind

In the last two sections we have seen how physics has often seemed to be in opposition to those who wish to defend the significance, even perhaps the primacy, of mental events. Now we shall suggest some reasons why the opposite view might, in fact, be more realistic.

First, however, I should emphasise my belief that some knowledge of modern theoretical physics is a prerequisite for any serious attempt to study the issues involved in the problem of consciousness. To avoid the insights which it can give is to unnecessarily handicap the investigation. Conscious mind, whatever it may be, exists within, and interacts with, the physical world. This is an inescapable fact.

The aspect of physics which seems most likely to have real relevance to consciousness is quantum theory. As we shall find in chapters 10 and 11, an essential feature of quantum theory is that the basic laws of physics, as stated for example in text-books, do not tell us what *is*, but instead they tell us about what will happen when we *make an observation*. In fact they only do this in a probabilistic way, i.e. they tell us the probability of certain outcomes of the observation. The theory does not, however, specify precisely what is meant by the phrase "making an observation". The only certain thing is that objects which are themselves described by the theory *cannot* make observations; *something else* is needed. Many people have argued that this *something else* is, or is related to, consciousness. As a very recent example, on my desk as I write these words is a document from the world-renowned Lawrence laboratory at Berkeley entitled *Quantum Theory of Consciousness* (Stapp, 1989b).

All this will be discussed in more detail later, and here we merely note two things. Firstly, the key point of quantum theory is not the lack of determinism, but the peculiar nature of reality which it suggests. Secondly, quantum theory implies that physics is much more open to what are sometimes regarded as "unscientific" ideas like consciousness, holism, etc, than it was considered to be in the last century, and than the less "reductionist" sciences like molecular biology and neurophysiology. There is an interesting irony here. In the previous sections we have noted a general feeling that the laws of classical (which normally is taken to mean non-quantum) physics were somehow too "simple" to contain the richness of conscious experience. Remarkably, however, through taking things apart and studying the microscopic constituents of matter, i.e. through reductionism, we have discovered a world which is far more mysterious than anything for which the world of normal experience had prepared us, and which we seem to find impossible to imagine or even describe. Perhaps instead of wondering how the laws of physics can lead to the amazing phenomenon of consciousness, we should be asking why the experiences of the conscious mind are so inadequate to permit us to understand even the behaviour of, so-called, "elementary" particles. In a recent defence of materialism I read *... brains are still machines, governed by the straightforward laws ...* (Friday, 1988). Such "straightforward laws" do not exist.

Reductionism, then, may not necessarily be the blind alley that leads to meaninglessness, it may rather be the way that opens out to a vista that far extends the domain of what we have previously thought of as "scientific". Certainly, even a minimal acquaintance with the mystery of the microscopic world should make us humble in our claims to *understand* anything. It should also make us willing to believe that nature may well have many more surprises awaiting our attention.

We shall see that there are some remarkable similarities in the problem of what causes "observations", in the quantum mechanical sense, and what gives rise to consciousness, so it is natural to think that the two issues may be related. Note, however, that *two problems* do not make *one solution*. This point is worth making because sometimes authors seem to say that we should not worry about not understanding conscious mind because we do not understand quantum theory either. We should at least *try* to do better.

The basic problem of quantum theory is that there is a gap between what the theory says about the world, and the experience we have of the world. There is another such gap in physics. This concerns time. All the fundamental laws of physics are unchanged if we reverse the directon of time, i.e. replace t by $-t$. (We shall have to qualify this statement in chapter 8 where we discuss time in more detail.) Such a symmetry in the direction of time, however, is clearly not a property of the world of our experience, not least because we remember the past and not the future! Why do the past and future appear to be so different, if the world is described by laws that do not distinguish between them? Is this also something to do with consciousness?

Another positive contribution which I believe physics can make to the study of consciousness is that, unlike the discipline of philosophy, physics is not hampered by a tendency to revere too much the authorities of the past. The subject tries to depend on empirical evidence and new evidence has continually overthrown cherished, and once thought indispensible, beliefs. It matters not who believed "law X"; if law X is violated by a carefully verified set of experiments, then law X is wrong! I think this is a useful antidote to some of the writing in philosophy, where ancient authorities are often quoted with great respect. An interesting example of this is that I have seen several articles where authors have been, to my mind, excessively concerned with whether their suggestions are consistent with the second law of thermodynamics. Quite apart from its problems with time reversal discussed in chapter 8, this is only an empirical law, which would readily be abandoned if and when we find circumstances where it fails. Similarly we would have no more respect for the law of conservation of energy, for Lorentz invariance (special relativity), three dimensions of space, etc. Only truth, by which here we mean the evidence of our senses, which we obtain through experiments, is sacred. Thus, when Eccles begins an interesting article (Eccles, 1986), to be mentioned again later, with the words: *All attempts to formulate a dualist hypothesis on brain–mind interaction are met with the strong criticism that such an hypothesis violates the conservation laws of physics*, the criticism to which he refers does not, I hope, come from physicists themselves. Similarly, Searle clearly believes that physicists would be unhappy to accept *"some entity that was capable of making molecules swerve from their paths"* (Searle, 1984, p.92). There are, of course, several such entities (we call them "forces"). Presumably he means that we would be unhappy to find any more. But why should this be? The history of physics (and of science in general) is all about the discovery of new things.

Finally, it is pertinent here to mention something that belongs not to physics but to the discipline of mathematics. An important theorem of mathematical logic, due to Gödel, seems to suggest that there is a concept of *truth* which goes beyond anything which can be proved by a series of well defined steps, used for example by a computer. Maybe even truth depends for its existence on conscious mind. We discuss this topic in chapter 9.

In this section we have been concerned mainly with the contribution that physics might make to the study of conscious mind. The contributions, however, are by no means all in one direction. Three possible places where an understanding of conscious mind might provide insight to physics are the measurement problem of quantum theory, the question of causation and "purpose" in the universe, and the whole area of aesthetic questions like what we mean by "simple" explanations, what are the proper questions to ask of a physical theory, etc.

Chapter 3

Consciousness

3.1 What is consciousness?

This book is about conscious mind, or consciousness, and we have
already used these terms many times. Alert readers will have no-
ticed that we have not defined them; we have implicitly assumed
that they would be understood, an assumption which is presumably valid for readers who have got this far, and which in general
is possible because consciousness is common to all people. This is
fortunate, because it does not seem to be possible to *define* consciousness in any meaningful way or to express it in terms of other
things. We have an excellent precedent here. *The Oxford Companion to the Mind* (Gregory, 1987) defines it as the *most obvious and
mysterious feature of our mind*, without needing to go any further.
Honderich (1988) gives a critical discussion of some attempts to
find a formal definition.

Consciousness is a type of perception. But it is not perception
of the external world; rather it is an individual's perception of his
own internal mental state. It is for each of us a private universe
of our own, where we are in control, where we can play all sorts of
imaginative games and can produce at will an apparently endless
variety of images, it is the origin and the home of all experiences
and feelings, it is the thing that somehow makes me *me*, essentially
distinguished from all other things and beings, it is the proof that
I am—not because I think, but because **I am aware**, i.e. because
I am conscious.

To my consciousness I have complete access, but, seemingly,
other people have access to it only through what I choose to reveal
by "physical" means. A child who does not like school quickly

realises the essential *privacy* of consciousness. *I am not feeling well* is a statement that cannot legitimately be challenged. We can test for a variety of possible causes of not feeling well, and we can look for symptoms of being unwell, but at the present time (and for all time?) it is impossible to see how a *feeling* could be confirmed or denied.

Many of the most significant "things" of our world would not exist if there was not such a thing as consciousness. "Red", for example, has no meaning outside of conscious mind. It is often caused by light of a certain wavelength, but red itself is not a wavelength. I have no way of measuring the sensation of redness, or even of knowing whether what I see as red is the same as what another person sees. The concept of colour is of course just one example of the many things that *hold their residence solely in the sensitive body.* (These words of Galileo are quoted, and supported, in a recent article by Boghossian and Velleman, 1989.) The same is true of the whole range of what we call emotions or feelings; of love and hate, of pain and pleasure, of envy and compassion. These are the essential ingredients of our lives; they are the things about which we care the most; they are attributes or properties of conscious mind.

There are other things which, less obviously, are, or might be, properties of conscious mind: free-will, the flow of time, perhaps even the concept of truth, maybe even the universe itself, at least in the form we know it. All these topics will be discussed in later chapters.

We return to the title of this section: What is consciousness? So far we have interpreted this as what do we mean by the word "consciousness"? In the same way, if we were asked what is a chair, we could say that it was something to sit on, or we could point to specific examples. However there is another way of interpreting the question about a chair; not what do we mean by the word "chair", but what *is* it. The answer might be pieces of wood put together in a particular way, or, ultimately, quarks and leptons suitably arranged. In the same way we could ask of consciousness, what *is* it? Does such a question make sense? It is presumably reasonable to say that brains are made of elementary particles, but is consciousness? Is it *made* of anything? Alternatively should we say that it is like a genuine elementary particle, for which the question "What is it?" would not have any meaning. A truly elementary particle is not anything other than itself. In our present state of knowledge a quark, for example, is not made of anything it just *is*. We should be careful with this analogy, however, because all quarks of a given type (see section 4.4) are identical. This is not true of conscious minds; mine, for example, is clearly different to yours. If we suppose that each is a different elementary object, then we have the very strange picture that there has to be lots of them waiting around to acquire names as people are born! The opposite extreme would be to say that there is only one consciousness, and that "my" consciousness is just a particular part of it.

None of these various suggestions appear convincing. Are there others? Or are the questions themselves somehow not permitted? But then, why not?

3.2   What is conscious?

Many years ago I was at a meeting in Didcot where the problem of whether fishing is cruel to fish was discussed. Naturally the issue depended crucially on whether fish felt pain, more generally on whether they were able to *feel* anything, i.e. upon whether they were conscious. As a very young scientist (it was indeed *many* years ago), my simplistic reaction was that such an issue was not one to be left to the uncertainties of people's opinions; clearly it should be answered by proper scientific investigation. Of course this quickly led me to the important and well known impasse, which to some degree has troubled me for over thirty years, that *it is not possible to devise a test that will tell us whether anything is or is not conscious.* The basic reason for this is of course quite simple; how *can* we test for something when we do not know what it is?

Let us see how we might try to decide who, or what, is conscious. I can begin with the fact that I am conscious. This I *know*. In one sense it is *all* I know. It therefore seems reasonable for me to extrapolate (from one example!) to the conclusion that other people are conscious. Indeed, as we have already noted, the fact that *you* have read this far presumably means that you have some idea what this book is supposed to be about, which in turn requires you to be conscious. This argument however is somewhat circular. In principle you could be reading the book without having any idea what it is about, just as a computer might read some given input regardless of whether it meant anything. It is essentially because I believe that you are a conscious being, that I am able to assert that you would not bother to read a book the contents of which were totally meaningless to you. Thus, basically, it is only by analogy with myself, i.e. because you, and other people, are so like me in your appearance, habits and behaviour, that I have any reason to believe that you are conscious.

The only doubts that I have seen expressed about the universality of human consciousness are due to a Princeton psychiatrist, H Janes. In a book, *The Origin of Consciousness in the Breakdown of the Bicameral Mind* (Janes, 1976), he claims that consciousness developed in people around two to three thousand years ago, mainly in response to considerable upheaval in their society. Previous to these upheavals they had acted in response to "voices" in their minds, and were unaware of any sense of responsibility for their behaviour. The relics of these voices still exist in cases of schizophrenia. Janes defends this claim by a detailed, and fascinating, study of the art and the literature of what he would regard as a preconscious age.

Here we shall follow the more widely accepted opinion that people are normally conscious. Of course, other people are not *identical* to me, so I am using some sense of being "approximately" like me, in order to extrapolate from the fact that I am conscious to the expectation that other people are also. But how approximate am I prepared to allow? Do I admit that animals are conscious? Clearly in many circumstances they behave in ways that are remarkably similar to the ways people behave. In particular, we note that this is true not only for what we normally regard as unconscious actions, but also for behaviour which we naturally attribute, in the case of humans, to conscious decisions. Nevertheless, such behaviour is not necessarily a signal of consciousness, as we can see by a trivial example.

If I am in a heated room and it becomes too hot then I will turn off the heat. A cat, on the other hand, might begin to look uncomfortable, but it would not take any action to turn off the heat. We would attribute this either to the fact that the cat only appeared to be uncomfortable, but actually did not "feel" anything and so was unmotivated to do anything about it, or, more realistically, to the fact that it did not know the whereabouts, or purpose, of the heating switch. Contrast now the cat's behaviour with that of a

thermostat. The thermostat would respond to the rising temperature in the same way that I do, i.e. it would turn off the heat. In
spite of this, I am sure we would be reluctant to regard the cat as
being less likely to be conscious than the thermostat.

Of course we could claim that this example is too specialised,
and point out that, in most circumstances, the cat's behaviour is
indeed more like that of a person than is the thermostat's. However, this would be being unfair to the thermostat, which only has
a very limited range of possible actions. Indeed it has only one,
and with regard to that degree of freedom, it behaves exactly as a
person.

Do we then assert that a thermostat is conscious? I suppose the
majority of us would be very reluctant to do this. Rather we would
say that the thermostat was "designed" explicitly to behave as it
did, and we would therefore attribute its behaviour, not so much
to its own consciousness, but to the consciousness of the person
who designed it.

Other opinions are, however, held. John McCarthy, the orig-
inator of the phrase "artificial intelligence" (or AI), claims that
... *machines as simple as thermostats can be said to have beliefs.*
This statement, which I have taken from John Searle's book *Minds,
Brains and Science* (Searle, 1984), is surely claiming that thermostats are conscious! It is interesting to note that, if we accept
this, then there are several possible ways of understanding the behaviour of the thermostat. It could for example turn the heat off
because

(i) it felt too hot, or

(ii) it wanted to please me, or

(iii) it was afraid that if it did not switch the heat off I would
throw it away and get a new one.

The more I try to understand what any of these imply, the more
absurd they appear.

There is, however, one further way of describing the behaviour
of the thermostat in terms of its being conscious. We could suppose
that *everything* is conscious, maybe at the same time introducing
the idea of degrees of consciousness so that some things could be
"less conscious" than others. In this case we might want to suggest
that an electron, say, moves in an electric field because it *wants* to
get closer to the positive charge (or whatever is causing the field).
Similarly all the laws of physics would represent preferences chosen

by the consciousnesses of the objects concerned. The thermostat would then also be behaving in this way but it would be choosing to obey the laws of physics, not choosing to turn the heat off. The fact that, in effect, it *did* turn the heat off would be a consequence of its design, i.e. again ultimately of the consciousness of its creator.

We shall return later to the idea of everything being conscious (section 5.8). One argument against it is that I know of one thing that is not conscious, namely, myself when I am asleep, or, even better, when I am under a general anaesthetic. A few years ago, I had a piece of cartilage removed from my knee. Throughout the operation I was a living, breathing, human being, but I did not feel the surgeon's knife. This gives me some confidence that when I break a stone, the stone is not hurt; that when I scrap a car, I am not being cruel; even that in plucking a flower, whilst I might be spoiling the scene for others, I am not spoiling the flower's happiness. But how far can I go? As the objects become more "like us" our doubts grow. But there is no obvious discontinuity. It is hard to see why we should make a sharp separation between one species and another, e.g. dogs are not conscious but cats are. Maybe this again is pointing to the idea of degrees of consciousness.

We are back to where this section began. As the earlier example showed, the question has great significance because it has at least the potential for affecting our behaviour. So far, however, we have made no progress towards finding an answer. Later, in chapter 6, we shall discuss some specific experiments designed to throw some light on the problem. It will be mentioned again several times in our subsequent discussion.

3.3 Can a machine be conscious?

Rather than asking whether given *objects* are conscious, we shall now ask whether it is possible to *construct* an object that is conscious, i.e. whether a machine, defined to be something that we are able to design and make, can be conscious.

Clearly we must discuss this issue under the assumption that not everything is conscious, since otherwise the question is pointless. It seems to me that it then is a very easy question to answer, and that the answer is a straightforward "no". The reason is very simple. *I do not know how to make a conscious machine because I do not know what consciousness is.* How can I possibly make something if I do not know what I am supposed to be making, and have no means of recognising it even if I succeed?

Two qualifications are needed here. First, I may succeed in making a conscious machine "by accident". It may just happen that I hit upon the correct ingredients and so, without knowing how or why, manage to make something that is conscious. I am grateful to a reviewer of an early version of this book for pointing out one familiar example of something like this: every baby is a conscious being created by its mother. Such cases however are clearly different to our actually designing and creating something which is conscious. The second qualification is that I am not saying that it is, *in principle*, impossible to make a conscious machine. One day we may learn what consciousness is, and then the argument is no longer relevant. It does apply however at the present time. Until I know what consciousness is, or what are the features necessary to produce it, I have no idea how to even begin to make a conscious machine.

To understand this a little better, let us suppose that a person from a society that had no written language, and who was therefore totally unacquainted with the concept of writing, was asked to make a book. In order that he might have some idea how to do this he was taken to a library, and allowed to see lots of examples of things that were books. However, being unable to read, he would not learn anything about what books really are. Hence, although he could make something that looked, superficially, like a book, and that felt like a book, etc, he could not make a real book. A vital ingredient, indeed the key ingredient, would be missing. However carefully he tried to copy the form of the binding, the stitching etc, he would not be able to make a book until he knew what a book was, i.e. knew about the meaning of language.

We must now carefully distinguish the above problem from one that is superficially similar, but is in fact very different, namely, that of making a machine that *appears to be conscious*. This is, of course, very easy. (It has always been easy in principle; recent developments in computer technology have made it also reasonably easy in practice.)

To be precise, what this means is that if you specify, *in advance*, the tests that you are going to apply to see if my machine is conscious, i.e. if you tell me what you are going to mean by being conscious, then I can guarantee that the machine I make will pass them, and hence will convince you that it is conscious. Thus, for example, I could make my robot "cry" when you spoke roughly to it, I could make it seek "food" (say a source of electric power) when it was "hungry" (its batteries were running down), I could make it complain if it saw that other robots were being given less work, or more honour, etc. All that we are doing here of course is finding ways of making the machine tell you that it is conscious. The easiest way of doing this is to put in a piece of magnetic tape that allowed the machine to say, loud and clear, *I am conscious.* Such a machine might fool many into believing it. It would not however fool us, because we know that we could just as easily put in a tape that said *I am not conscious.* Neither reply would have any bearing whatever on the question we are trying to answer. The sad thing is that the tests we are applying are not simply unreliable; they are irrelevant! Incidentally nothing here depends on whether or not the robot is, or is not, *actually* conscious.

The point of the above paragraph has been much emphasised by J Searle. In his language what we are doing with our robot is "simulating" the effects of consciousness, and there is no necessary reason to believe that by simulating the effects of consciousness we are creating it. It is worth quoting again his way of illustrating this distinction (Searle, 1987, pp.213, 214):

"Suppose I am locked in a room. In this room there are two big bushel baskets full of Chinese symbols, together with a rule book in English for matching Chinese symbols from one basket against Chinese symbols from the other basket. The rules say such things as 'reach into basket 1 and take out a squiggle–squiggle sign, and go put that over next to the squoggle–squoggle sign that you take from basket 2'. Just to look ahead a moment, this is called a 'computational rule defined over purely formal elements'. Now let us suppose that the people outside the room send in more Chinese symbols together with more rules for shuffling and matching the symbols. But this time they also give me rules for passing back Chinese symbols to them So, there I am in my Chinese room, shuffling these symbols around; symbols are coming in, and I am passing symbols out according to the rule book. Now, unknown to me, the people who are organising all of this on the outside of the room call the first basket 'a restaurant script', the second basket 'a story about the restaurant'; the third batch of symbols they call 'questions about the story', and the symbols I give back to them they call 'answers to the questions'. The rule book they call 'the programme', themselves they call 'the programmers', and me they call 'the computer'. Now after a while, suppose I get so good at answering these questions in Chinese that my answers are indistinguishable from those of a native Chinese speaker. All the same, there is an important point that needs to be emphasised. I don't understand a word of Chinese, and there is no way that I could come to understand Chinese from instantiating a computer programme in the way that I described it. And this is the point of the story: *if I don't understand Chinese in that situation, then neither does any other digital computer solely in virtue of being an appropriately programmed computer, because no digital computer solely in virtue of its being a digital computer has anything that I don't have.* All that a digital computer has, by definition, is the instantiation of a formal computer programme. But since I am instantiating the programme, since we are supposing that we have the right programme with the right inputs and outputs, and I don't understand any Chinese, then there is no way any other digital computer, *solely in virtue of instantiating the programme* could understand Chinese.

"Now that is the heart of the argument. But the point of the argument, I think, has been lost in a lot of the subsequent literature developed around this, so I want to emphasise the point of it. The point of the argument is not that somehow or other we have an 'intuition' that I don't understand Chinese, that I find myself *inclined to say* that I don't understand Chinese but, who knows, perhaps I really do. That is not the point. The point of the story is to remind us of a conceptual truth that we knew all along; namely, that there is a distinction between manipulating the syntactical elements of languages and actually understanding the language at a semantic level. What is lost in the AI *simulation of* cognitive behaviour is the distinction between syntax and semantics".

A crucial distinction is being made here between being able to make formal preprogrammed responses to a certain input, and actually *understanding* the input. It seems to be a valid distinction, although it is not too clear what "understanding" really means. It is probable that understanding is also a property of conscious mind, so the argument begins to look a little convoluted! We could perhaps try to see the difference by noting that the robot, responding only by formal steps, would be totally incapable of making a proper response to a sufficiently clever question, whereas there seems to be a more "open-ended" quality in the ability to respond of someone who understands. I think this is reasonable, but wonder if I am being misled by the fact that the brain is such an incredibly clever computer.

In any case, it is beyond question that it is easier to make a machine to simulate consciousness, than it is to make one that *is* conscious. As we have already stated it is very hard to see how we can ever make anything (except by accident) if we do not know what it is.

Conversely, of course, if we do know what something is, then we might expect to be able to make it, indeed to make it in several different ways. Suppose, for example, that a particular set of electrical connections (clearly a very complicated set!) were necessary and sufficient for consciousness, then, although in nature such a set only occurs in biological material (e.g. in brains), there is no reason why we could not make it out of anything, bits of wire, even old tin cans, etc. There may of course be serious *practical* difficulties, but not any absolute, in principle, obstacles. I make this point because later in the above article Searle appears as though he is claiming something unique about biological material.

Finally in this section we note that ability to calculate, i.e. do formal steps, very rapidly, is not necessarily in any sense connected with consciousness. There are machines readily available that could beat me at chess (somewhat more expensive machines could beat most people). That is, such machines could consider the outcomes of various moves so quickly that they could always select sufficiently good moves to ensure that ultimately they would win. But, whereas I might feel sorry to be beaten, they would find no pleasure in winning and no sense of disappointment if they were to meet a better player than me and hence lose. Contrary to popular myth, not to care about winning or losing is to be a very poor "sport", and in this sense a computer is a total failure!

3.4   What does consciousness do?

In the first section of this chapter we asked what consciousness *is*, and did not have much success in finding an answer. Here we shall try a different approach and ask instead what does consciousness *do*, or, in other words, what is its effect upon the rest of the (non-conscious) world? Possible answers seem to range from essentially nothing, i.e. everything in the world would be pretty much the same even if we were not conscious, to essentially everything, i.e. the world as we know it is largely a consequence of the existence of consciousness.

The answer which most immediately comes to mind is inter-mediate between these extremes: conscious mind, we would be inclined to say, certainly affects us as individuals; at least in some circumstances our behaviour is controlled by our conscious mind. Whilst this answer may be true, there are at least three reasons why we should be cautious about it.

Firstly, it is obvious that much of what we do, even processes which are known to be controlled by the brain, we do unconsciously, i.e. the conscious mind plays no role. This is clear for things like breathing, blinking, digesting our food, adjusting our body tem-perature, etc. In addition there are many skills, which we may learn consciously, but which we soon come to use unconsciously, e.g. driving a car, skiing, writing, etc. Even the act of withdrawing our hand, for example, from something that is hot, although we might explain it as not wishing to be burned, is in fact performed without our consciousness having time to "think about it". Indeed there are many cases in which conscious mind is a hindrance to effective action; we perform better without it. A simple example here is that I do not consciously think about the positions of the letters on the keyboard as I type this book. I could not write down the pattern of the keyboard, but it is a fact that, unconsciously, my fingers tend to go to the right places to find the letters. Conscious-ness in general seems to play little role in normal circumstances. It is when things go wrong, or when novel circumstances occur, that we call in consciousness.

Secondly, even if we admit that there is something, e.g. a region of the brain, or maybe something totally different, that exerts a controlling influence over our actions when needed, why does it have to be *conscious*? What, if anything, is there about consciousness itself that affects our behaviour? Does the fact that I am *conscious*, as distinct from the fact that I am clever, i.e. a superbly constructed computer, really make any difference. Of course it makes a difference to me, that is to my consciousness itself, and to my perception of the world. Without it there would be no such perception, and no such things as beauty or understanding; the world would be a dull place. Does the existence of these things, however, actually alter this outside, physical, world? Do they for example have any survival value so that they might have been expected to evolve in a world that did not, initially, contain them?

Thirdly, it is evident, as we have already noted, that the behaviour of animals and humans is in many ways remarkably similar. If, following Descartes (section 5.6), we believe that animals are not conscious, then it would seem to be necessary to conclude that consciousness is not responsible for the behaviour of humans where that behaviour is similar to that of animals. Its role would then be limited to explaining the difference in the behaviour of humans and animals. Of course, everything in this paragraph is dependent on the assumption that animals are not conscious.

The above points show that we must be careful in naively accepting the view that our behaviour is in the control of our conscious mind. Much of what we do is independent of the fact that we are conscious, and it is possible to envisage non-conscious beings with very similar behaviour patterns. Nevertheless, it is important to state clearly and confidently that the physical world is affected by the existence of consciousness, i.e. if there was not such a thing as consciousness the world would be different. The "proof" (of course in this game there is no such thing as a proof in the way that a mathematician would use the word) of this lies in the fact that it is inconceivable that I would be writing this book if I were not conscious. Or, even better, we can surely agree that, if nobody was or ever had been conscious, then the word itself would not appear in dictionaries. Since it does, and since dictionaries are physical things, I think we can regard the result as established: the existence of consciousness has an effect on the physical world. There is no doubt that in trying to understand the physical world we are forced into having to accept some very implausible things (as we shall see in detail later), but to some standards we must keep, and this result is one of them. *How* consciousness has this effect, and how significant it is, are not answered here. They are questions to which we shall return.

If we now allow that some things are not conscious, and follow the general ideas of evolution, then we would expect to find that some of the effects of consciousness have survival advantages. Presumably this is only possible to the extent that the existence of consciousness has a physical effect (survival is, after all, something physical). Provided we allow this possibility, it is in fact not difficult to see how consciousness could convey such advantages. As we saw in the last section, we can easily programme a robot to "look after itself" in a variety of ways; seek a source of electricity when its batteries are low, or shelter when it begins to rain, etc. We would of course have to anticipate the hazards in order to programme the machine suitably. How much easier, and more reliable, would this be if instead we could somehow programme the robot to *want* to survive. Then it could use its vast computer skill to find the best way of doing this. Such a programme would be impossible unless the computer understood the concept of wanting. This requires it to be conscious. It is interesting to note that biologists have claimed that even very primitive life-forms can behave in ways that makes them appear as though they also possess some sort of will to survive. If it was possible to establish that this was not just "appearance", but was genuine, then we would indeed have an exciting result. However, for reasons that we have already given, there does not seem to be any way by which this might be done.

It seems to me that the idea of "wanting to survive" is a more significant advantage of consciousness than the only other one that I have seen suggested, namely, that consciousness helps with social relationships because it allows a person to put himself into the position of another, and hence guess what the other is thinking (Humphrey, 1986). Weiskrantz (1987, p.311), indeed, challenges whether this is in fact an advantage, pointing out rather that humans are the only animals that get into social difficulties of any sort, and that this might be due to our tendency to bother too much what others are thinking!

3.5   The uniqueness of conscious mind

There is another important role that conscious mind appears to play in our life, namely, that of providing some sort of continuity and unification of our experiences. The continuity in time occurs for example in our appreciation of music; we somehow put separate notes together to make a tune. On a much longer time scale, conscious mind readily jumps gaps of unconsciousness, e.g. in sleep. It will be the same "me" occupying my body tommorow as it is today; hence I worry about having to go to the dentist next week. I do not understand the claims of Parfit (1987) which would seem to suggest the contrary:

> *Suppose I know that tomorrow I shall be in pain. If I knew that, after my death, a Replica of me would be in pain, I would not fearfully anticipate this pain. And my relation to myself tomorrow is no closer than to my Replica.*

This does not accord with our experience, and I think we would take a lot of convincing! Of course, in so far as conscious mind is part of my physical body, there is no problem in understanding why there is continuity between me today and me tommorow; for the most part the actual atoms are the same.

A similar sort of effect occurs spatially. The different parts of a picture presumably cause physical events to happen at distinct sites in the brain. The picture, however, is seen as one thing. To some extent of course we could say this is just an example of how clever the brain is at interpreting information. There does, however, seem to be something more than this. We do not *experience* different parts of the brain. Consciousness insists on being *one thing*. We do not ever have any sense of it being made of different parts which communicate with each other. If consciousness is somehow a property of very complex pieces of matter, then we might expect that various parts of the brain should be conscious by themselves. This, however, is not how we experience consciousness. My consciousness is a unique "something". This does not conclusively exclude the possibility that it is made from different "parts". These may be so intimately connected that I am not aware of their separateness.

3.6 The unconscious mind

We have already noted that our brains are involved in a large amount of activity of which we are totally unaware. They are then, presumably, acting as non-conscious machines, just like extremely powerful computers. Except for their quite awesome power, there is nothing very mysterious about this. The mystery is in consciousness, not lack of consciousness. It seems therefore to be unfortunate that psychologists talk about the "sub-conscious", and about "unconscious desires", etc, as though these existed at some deeper or more mysterious level than the things of consciousness.

What we call the sub-conscious is simply the brain acting as any other non-conscious machine. The Freudian idea that unconscious desires have been pushed out of the conscious mind by a sort of censorship is picturesque, but does not seem to be necessary. Surely nothing more than the causal laws of physics need be operating here (see the remarks of Russell, 1921, p.38). If we think, for example, about sexual desire, there is nothing very surprising about the fact that the sight of a young woman might cause certain processes to occur in the neurons in my brain, which in turn could trigger responses elsewhere. This can all be seen as the consequence of the need to survive, and hence to procreate, through millions of years of evolution. The really *difficult* thing to understand is that I can become aware of this, that I can imagine the pleasure the said lady might give, and that I can think about the morality of any proposed action. It is the *conscious* mind that is most in need of understanding.

We must, however, insert a qualification here. When Freud, for example, spoke of the sub-conscious, he was probably working with a picture of a person as two separate entities: a physical body, including the brain, and a "mind". Then the latter could be thought of as working at two levels, a conscious level, and a sub-conscious one (see figure 3.1). It is possible that this sort of thing is true. If we accept the ideas of dualism, i.e. that there is a real separation between the mind and the brain (see section 5.6), then we could imagine that the non-physical mind itself has some sort of structure, allowing it to operate at both a conscious and an unconscious level. Swinburne (1987) gives a detailed discussion of such a structured mind. However, it seems to me that we are here entering into an unnecessary complication. All the non-conscious functions of the mind can perfectly well be understood as the operation of physical machines; it is only when we reach the phenomena of consciousness that we begin to wonder about whether this remains true. This is the view we shall adopt throughout the remainder of

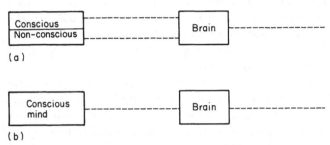

Figure 3.1. (a) Here we picture a structured, non-physical, mind, which can operate in either a conscious, or non-conscious, manner.
(b) Here all non-conscious processes are seen as physical processes in the brain. The question of whether the non-conscious mind processes are physical is left open.

this book.

3.7   States of consciousness

There is a long tradition which speaks of the possibility of people, through contemplation and training, being able to attain to states of "higher" degrees of consciousness, even indeed of "pure" consciousness. Ideas of this sort are particularly associated with mystics and with followers of Eastern religions. At a very basic level it is clear that we can all cut ourselves off from most outside influences and concentrate upon our minds, thereby decreasing our awareness of any sort of sense input, and at the same time increasing our degree of awareness of our own thoughts. Is this what is meant by a state of pure consciousness? Brian Josephson, who was awarded the Nobel Prize for his discovery of the so-called Josephson effect, and who is now professor of physics at the Cavendish Laboratory in Cambridge, writes:

> It is a familiar feature of physical systems that there exist some systems which can be described completely (with reference to some particular point of view) in a simple way. Examples are the ground state of liquid helium, or a perfect crystal of sodium chloride at a given temperature. This may be contrasted with the situation for chemically impure

substances or disordered systems. In the case of conscious experience we assert that the same situation obtains, that is that simply specifiable states of consciousness exist. Typically, these states consist of what may be called "pure" ideas or emotions. Most basic of all is the state known as pure consciousness or samadhi, which has no identifiable content other than being conscious. This may be understood theoretically in terms of the following picture. Pure consciousness is that limiting state of consciousness which is completely undisturbed by other entities; in other words it consists only of the phenomenon of consciousness interacting with itself (Josephson, 1984).

In principle, it appears likely that the study of consciousness through these types of activity could help us in understanding its true nature. On the other hand, it could be that, by deliberately excluding the obviously physical influences, we are making it even harder to understand the relation between consciousness and the physical world. This whole topic is, to me, mysterious and obscure, so we shall ignore it in what follows. It just may be that in so doing we make a big mistake.

There does exist one very familiar state which is a sort of "pure" consciousness, namely, that of dreaming. The things we are aware of when we dream exist *only* in our minds (that at least is normally the case). Of course, the people and objects that appear in our dreams may *also* exist in the "real world", but the particular actions about which we dream have no such reality. How then do dreams differ from merely imagining particular events?

If, or when, we ever have a science of conscious mind, there is little doubt that states of contemplation and of dreaming, etc, will play a big part in the experiments we do. Maybe then we will understand them better than we do at present.

This chapter has been very vague, and it has contained a lot of questions with very few answers. With much relief therefore we return, in the next chapter, to the world of physics, where some things are, apparently, much clearer.

Chapter 4

From classical physics to the standard model

In this chapter we shall give a rapid review of the remarkable success story of modern physics, which is mainly the physics of the present century. This will help us to see what is meant by an "explanation" in physics; it will show how new facts and new understandings often require quite radical changes in attitude; it should also encourage us to believe that observable phenomena *can* be understood; finally, although much of this chapter may seem to be irrelevant to our purpose in this book, it is necessary in order to place quantum theory in its proper context.

I have written an expanded version of this story in *To Acknowledge the Wonder* (Squires, 1985). Other popular accounts are given in Close (1983) and in Pagels (1983). Readable books on related material, including some cosmology, are Barrow and Silk (1983), Hawking (1988) and Weinberg (1977).

4.1   The classical era

In order fully to appreciate the revolution brought about by quantum theory, and the great advances that have recently been made, we shall first consider the state of physics around the end of the last century. The physicists of that time had seen a great deal of progress; by careful experiments and through the use of mathematical analysis, they had come to understand many natural phenomena; they were confident, perhaps arrogantly so; and some probably thought that almost everything that was capable of being understood, was understood. From the vantage point of today, we can see how very restricted was their knowledge, and how very limited their ambition!

Scientists of the 19th century knew, or at least thought they knew, of three types of "object": particles, forces and waves (these are not really independent entities but it is convenient here to treat them separately). We shall consider them in turn.

Particles

The particles were principally the atoms and molecules of which all matter is made. Ninety-two different types of atom, corresponding to the ninety-two different elements, e.g. hydrogen, helium, oxygen, iron, lead, uranium, etc, were known. These atoms fitted into the pretty, but mysterious, "periodic table", so called because, when elements were arranged in order of the masses of the corresponding atoms, those with similar properties occurred at intervals. All the properties of the atoms, however, were both unexplained and inexplicable; atoms were the "given" stuff of the universe. As we have already noted, no less an authority than James Clark Maxwell had written that the origin of the properties of atoms would be, *forever*, beyond physics!

Most material substances are not elements, but are compounds, in which the constituent particles are molecules rather than atoms. Each molecule is made from a precisely defined number of, in general, different types of atom. The properties of the molecule are usually very different to those of the constituent atoms. For example, a water molecule contains atoms of hydrogen and oxygen, both of which are gases, whilst the molecule of common salt contains an atom of chlorine, a highly toxic gas, and an atom of sodium, a metal which spontaneously ignites in air. Again, to the scientists of the last century, there was no hope of explaining why the atoms bound together to form molecules, or how these molecules acquired their properties.

The atoms and molecules were the particles involved in the kinetic theory of gases (statistical mechanics), which had brilliantly, and contrary to the expectations of some scientists (see section 5.2), explained the laws of thermodynamics and, for example, the fact that the volume of a gas is inversely proportional to its pressure (Boyle's law).

One other particle, not an atom, was also known. This was the negatively charged "electron", which could be pulled out of atoms and molecules by sufficiently strong electric forces. It was natural therefore to suppose that atoms, which are electrically neutral, are some positvely charged "stuff" containing isolated electrons.

Forces

Since the time of Isaac Newton (1642–1727), forces have been known to be "things" that changed the state of rest or uniform motion (i.e. constant velocity) of material bodies. Two fundamental examples had been found. The first was the one Newton himself had postulated, namely, the gravity force between all bodies. This force is always attractive (i.e. it pulls objects towards each other); it is proportional to the product of the masses of the two objects (hence the fact that all bodies fall at the same speed in the pull of the earth's gravitational force); and it reduces as the square of the distance between the objects (since this is in fact a rather slow rate of reduction we refer to gravity as a force of infinite range). It was by using this force law that Newton, in 1680, had explained the motion of the earth and other planets around the sun, and thereby had begun the process of bringing "the heavens" within the domain of physics.

The second set of forces were those associated with electricity and magnetism. The simplest manifestation of these is in the electric force between two charged objects. Again this falls off as the square of the separation, but this time the strength is proportional to the product of the *charges*, rather than the masses. Magnetic forces were also known, and careful experimental work by Faraday and others had led Maxwell, in 1864, to the theory of electromagnetism, which unified all electric and magnetic phenomena. As we shall see below, Maxwell's theory also predicted many totally new effects.

Other forces, like the forces of contact between material bodies, the forces of tension in springs, etc, are not really new. They are just complicated manifestations of electric forces. This fact is in stark contrast to what was accepted prior to the work of Newton on gravity. Then it was believed that all forces were simply the "push/pull" effects of material bodies. Indeed Newton was embarrassed by the fact that his law of gravity seemed to suggest that action-at-a-distance was taking place: the presence of an object at one place could influence the behaviour of another an arbitrarily large distance away. "Materialism", in its narrowest interpretation, died in the 17th century.

Waves

These are the third ingredient of classical physics, although they are not really independent of the others. They come in two types, which need to be carefully distinguished. The first are the waves in material bodies, e.g. waves on a stretched string, or a taut membrane, ripples on the surface of water, sound waves, which are pressure variations in air or some other material, etc. These are all, in principle, straightforward consequences of Newton's laws of motion, together with a few elementary properties of materials, e.g. Hooke's law which tells us that the restoring force in an elastic body is proportional to the displacement from the equilibrium position. In the language of section 2.3, they are not really *new*, but can be *reduced* to the motion of particles acting under the influence of forces.

The other type of waves are crucially different. These are the waves of electromagnetism, of which light is the most familiar example. That visible light is a wave motion had been known since the beginning of the 19th century, mainly due to the work of Young on interference effects. The phenomenon of interference is so important to our discussion that we devote the next section to it.

The true nature of the waves which were responsible for visible light was not realised until Maxwell, by applying simple mathematical ideas, and some requirements of consistency, to the rules of electromagnetic interactions as deduced by experiment, predicted the existence of waves, travelling through space with a velocity that could be calculated from the strength of electric and magnetic forces. This velocity turned out to be identical to the measured velocity of light, 3×10^{-8} m s^{-1}. Thus, in a sense, light waves are also a consequence of the forces in nature. These electromagnetic waves are now known to embrace a large variety of phenomena (see table 4.1), depending on the wavelength (λ) or frequency (ν). As explained in figure 4.1 these quantities are related because their product always equals the velocity of light:

$$\lambda \times \nu = c. \qquad 4.1$$

Note that the velocity c is independent of the frequency.

Although electromagnetic waves are a consequence of what is often referred to as *classical* electromagnetism, to distinguish it

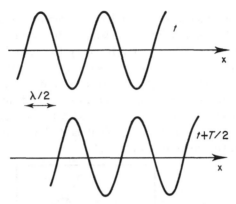

Figure 4.1. A simple wave at times t and $t + T/2$. At a fixed x the wave has completed half an oscillation, so $T/2$ is half the period, i.e. half the time for a complete oscillation. The figure shows that the wave has travelled a distance $\lambda/2$ and hence its velocity is $\lambda/T = \lambda\nu$ as in equation 4.1.

Table 4.1. Various types of electromagnetic radiation, and their approximate wavelength in metres.

name:	radio	VHF	micro-wave	infrared	visible	X-rays	γ-rays
λ:	10^3	10	10^{-2}	10^{-4}	5×10^{-7}	10^{-10}	10^{-17}

from the quantum version known as quantum electrodynamics, or QED, they actually represent a major break from the basic tradition of 19th century physics. That tradition had been to explain everything in *mechanical* terms, i.e. to reduce all things to the motion of particles. As we have seen, waves on strings, etc, were consistent with this; they were not really anything new. With electromagnetic waves, however, there was no obvious answer to the question: *What* is vibrating? Indeed we can describe a very simple experiment that shows the real significance of this question.

We imagine that we have a bell suspended inside a glass jar, as shown in figure 4.2. We ring the bell and we can both *see* and *hear* it. This is because light (electromagnetic) and sound waves travel to us from the bell. Now suppose that the jar is attached to a pump which can remove the air. As the air pressure inside the jar goes down, we notice a remarkable thing: although the sound of the bell diminishes, the visual image remains unaltered. The reduction in the sound level is easily understood. As the air is removed there is less in the jar to carry the sound waves. If we could reach the state of a perfect vacuum then we would not be able to hear the bell at all (though some sound waves would in fact be carried along the string that is used to suspend the bell). What, however, can we say about light waves? Clearly whatever carries these is still present in the jar. What is it that remains when we have, apparently, created a vacuum?

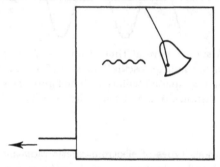

Figure 4.2. A bell inside a glass jar. When the air is removed we cannot hear the bell but we can still see it. What carries the light waves?

Maxwell's theory does not help here. Indeed the calculations make no reference to any medium carrying the waves; they really do move through "nothing". (In fact, it is possible to modify the equations to describe transmission through a medium. The resulting waves then have a different velocity.) In spite of this, the prevailing mechanistic view at the end of the last century required a medium to carry the waves, and the clear message of the equations was not believed. No known medium would serve the purpose so a new one, called the "ether", was invented. Because it filled all space, including apparent vacua, and for other less obvious reasons, it had to have very peculiar properties, and many were the contortions required to devise a suitable material. In the departmental library here in Durham we have handwritten notes of lectures given in 1884 by Sir William Thomson at John Hopkins University in Baltimore. The lectures were called *Molecular Dynamics*, though his subject was supposed to be the wave theory of light. That everything had a mechanical explanation was not in doubt:

> ... *we must not listen to any suggestion that we must look upon the luminiferous ether as an ideal way of putting the thing. A real matter between us and the remotest stars I believe there is, and that light consists of real motions of that matter.*

In trying to give an example of a suitable type of substance, he suggested Scottish shoemakers' wax!

In this book I have, and shall again, praise the merits of physicists and the methods of physics. The discussion here, however, illustrates the fact that serious errors have been made, in particular through an unwillingness to abandon prejudices. Scientists towards the end of the 19th century implicitly accepted what we might call a mechanistic materialism, the "molecular dynamics" of Thomson's lectures. This had been wonderfully successful in explaining thermodynamics—it must therefore also explain electromagnetism. They were unable to imagine that the world could contain anything else. They were wrong, and maybe here there is a moral for us.

We have now abandoned the ether, though it would probably be rash to say that it will never be revived in some form or other. The question of what is vibrating is one we do not ask. Whereas the quantity that varies in an oscillating string, for example, is the actual displacement of the string from its equilibrium position, the corresponding quantities in an electromagnetic wave, which are the electric and magnetic fields, are not displacements of anything, they just *are*. That is to say, we believe that these fields really do exist. At each point of space and time there is an electric and a magnetic field.

This picture of waves travelling through empty space seems to cause a new problem concerning their velocity. To understand this problem we first think about waves on a string. The velocity of these waves can be calculated (it depends on the properties of the string), and the number that we calculate gives the velocity *measured relative to the string*. Thus if we put the string on a moving train, the waves will move at a different speed relative to the stationary ground. Now return to electromagnetic waves. Here there is no medium which carries them, so what does the calculated velocity mean? A velocity has to be relative to something, i.e. we have to know what is at rest. Amazingly, the solution to all this was again to believe what Maxwell's equations actually said, namely, that any observer, regardless of his velocity, will observe the same value for the velocity of elctromagnetic waves. That this is contrary to our intuition is due to the fact that our normal experience does not involve such high velocities. The result, which was confirmed in the famous experiment of Michelson and Morley, described in figure 4.3, was the key ingredient of Einstein's theory of special relativity. This has since been checked in many different experiments, and is now an accepted part of the structure of physics.

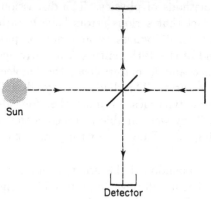

Figure 4.3. The Michelson–Morley experiment. The central mirror reflects half the light and transmits the other half. The two beams are brought together at the detector. If the light travels at different speeds in different directions, as would be expected because of the earth's velocity, then the interference pattern at the detector should change as the apparatus is rotated. This is not seen.

We shall not attempt to explain this theory here, but it is worth noting one very important feature, namely, that it encourages us to regard *time* as being very like a dimension of *space*. It is well known that we have freedom in choice of axes to describe positions, i.e. we can rotate the axes, and in special relativity this freedom is extended to include similar rotations among the time and space axes; there is nothing absolute about the particular choice of a "time axis". (See Squires, 1985, or any book on special relativity for more details.)

Finally, we mention a much studied form of electromagnetic radiation, namely, that which is emitted when elements are heated.

This consists of a set of (almost) monochromatic, i.e. single frequency, radiations, and hence gives narrow lines when the light is bent in a spectrograph (essentially a prism), which bends light through an angle depending on its frequency. These are the so-called *spectral lines*, and they are characteristic of the particular element which emits them. (Many readers will be familiar with the bright yellow line which every chemistry student knows signifies the presence of sodium.) The techniques of spectroscopy allowed the frequencies of the various spectral lines to be measured with great accuracy, a fact which, as we shall see, played an important role in the development of quantum theory.

4.2 Interference

In this section we digress in order to give a careful and very elementary discussion of the important phenomenon of *interference*. No proper appreciation of the significance, or of the problems, of quantum theory is possible unless interference is understood.

The idea is very simple and we can introduce it by considering any situation where there are several "sources" contributing to some total effect. For example, we consider putting fruits onto the pan of a balance. We know that if a plum weighs 100 g and an apple weighs 500 g, then both together will weigh 600 g. Having seen that the idea is totally trivial we shall put it into symbols. First, we weigh each fruit individually, obtaining a value w_i for the fruit labelled by i, with i going, say, from 1 to N, the number of fruits. (If we call the plum "1", then w_1 is 100 g.) The total weight when we put all the fruits on the pan is given by

$$W = w_1 + w_2 + w_3 + \ldots + w_N. \qquad 4.2$$

(Readers who do not like equations will be happy to note that this is how they obtained the value of 600 g in the illustration above.) Note especially that each of the fruits contributes a particular value, regardless of what other fruits are weighed. This is an example of what is usually called a *linear superposition principle*, a fancy name for something very simple. The w_i are, of course, all positive, so the addition of an extra fruit inevitably increases the weight.

As a second example, consider various contributions to a bank balance. If we denote these by c_i, then, assuming that the initial balance is zero, the total has the same form as the above, i.e. it is given by

$$T = c_1 + c_2 + c_3 + \ldots \qquad\qquad 4.3$$

There is now, however, a new feature because only some of the c_i are positive (salary, cheques paid in, etc) and the others are negative (withdrawals, standing orders, etc). Thus, including an additional contribution does not necessarily increase the total; it could instead reduce it, even reduce it to zero if the last contribution wipes out the effects of all the others. It is because this can happen that the phenomenon we are dealing with is called "interference". Interference occurs whenever we have a linear superposition principle in which some of the contributions can be negative. The name is not really a good one because the linear superposition principle still applies, i.e. the effect of one transaction is not in any way altered by the presence of the others, so the various contributions do not "interfere" in the sense of getting in each other's way. (Actually one could complicate things by introducing effects of this sort, e.g. we could include the possibility of my drawing out a larger amount if I saw that my previous total was larger than expected. Then the linear superposition principle would not hold, because a particular c_i would depend on the others.)

For the next example, consider the situation shown in figure 4.4 in which a beam of projectiles emanates from a point O, in random directions. Here, of course, we are thinking of the projectiles as *classical* objects, say, tennis balls. Some of the projectiles will pass through the holes in the barrier shown in the figure, perhaps after being deflected by the edges. The projectiles are then counted at various points on the detection screen as shown. We first do an experiment with one hole closed, and, after a given time T, record the number of projectiles at each point of the screen. Suppose this is $n_1(x)$, where the x specifies a particular point on the detection screen and the "1" signifies that this is the contribution of hole number 1. We repeat the experiment with the other hole closed, this time obtaining $n_2(x)$ for the number reaching x after a time T. Finally we keep both holes open. Then, provided that the rate of firing is such that there are no collisions and the projectiles do not get in each other's way (this is necessary for linear superposition to apply), and provided that T is sufficiently large so that the statistical fluctuations become negligible, we know that the number of projectiles reaching the screen will be the sum of those going through each hole separately:

$$N(x) = n_1(x) + n_2(x). \qquad 4.4$$

Clearly, at all points on the screen, N will be at least as large as both n_1 and n_2. (Notice that I asserted above that we "know" this result; we did not need to do an experiment with both holes open in order to write down the last equation. Readers may like to think *why* we were able to make this assertion, and what they would do if an experiment were to be performed which gave a different result.)

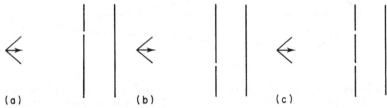

(a)   (b)   (c)

Figure 4.4. The contributions from the two slits separately, as in (a) and (b), give the total contribution expected in (c).

We now replace the source of classical particles by a source of waves. A simple, two-dimensional, example would be waves on the surface of a tank of water in which there is a barrier containing two holes. Each of the holes now represents a source of waves reaching any point beyond the barrier. Let the wave from the first hole at time t and position x be given by $h_1(x,t)$, and similarly from the second hole with h_1 replaced by h_2. Note that the x and t in the brackets are the standard way of expressing the fact that the quantities h depend upon the values of x and t. We use the letter h to denote the wave because it is appropriate to the example of water waves, where it refers to height above the equilibrium surface of the water. The same results however apply to other types of wave. The total wave, at position x and time t, is then given again by the linear superposition:

$$H(x,t) = h_1(x,t) + h_2(x,t). \qquad 4.5$$

Although this equation is similar to the previous one, its effect can be very different because now the h_i can be *negative*. Thus, in some cases, the wave from the two holes is less than that from either one hole separately. This is the phenomenon of interference of waves. It cannot occur with classical particles and is an unmistakable signal of the presence of waves. The actual mathematics of two-slit interference is explained in a little more detail in the caption to figure 4.5.

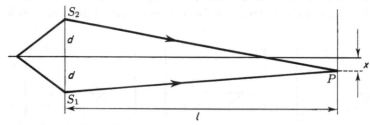

Figure 4.5. Two-slit interference. The difference in path lengths is $S_2P - S_1P = [(d+x)^2 + l^2] - [(d-x)^2 + l^2]$. There is complete destructive interference, i.e. zero intensity, when this is equal to $(n + 1/2)\lambda$, where n is any integer, and λ is the wavelength.

As we have already stated, it was the fact that light exhibited the phenomenon of interference that convinced physicists around the end of the 18th century that it was a wave motion. Since Newton had said that light consisted of particles, they took quite a lot of convincing, but ultimately there was no escaping the conclusion: *interference implies waves*.

Most of us, of course, will not have seen the two-slit interference pattern. One striking manifestation of the interference of light waves is, however, familiar to us all. This is the pattern of colours seen in light reflected from a thin film, e.g. oil on water. The two "sources" are the light reflected from the top and bottom surfaces of the oil, and the colours arise because different wavelengths, i.e. different colours, interfere destructively at different angles (see figure 4.6).

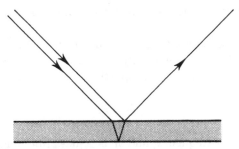

Figure 4.6. Light rays reflected from the two surfaces of a thin film have a different path length, so there is an interference pattern depending on the angle and the wavelength.

4.3   The quantum revolution

The quantum revolution began, in the early years of the present century, with the realisation that there is a fundamental conflict between two of the pillars of classical physics. To explain this conflict, we first need to introduce the idea of *degrees of freedom* of a system. These are the various independent motions of the system; for example, the velocities of the atoms of a gas in each of three mutually perpendicular directions (a velocity in any other direction is not independent because it can be expressed as a sum of these), the rotations of the atoms, the oscillation of the atoms in a two-atom molecule, etc. Next, we need to know that classical statistical mechanics, which results from the laws of mechanics applied to systems with many degrees of freedom and which forms the basis of the well established laws of thermodynamics, requires that the energy of a system distributes itself, on average, equally over all degrees of freedom. Since, roughly speaking, the various frequencies of electromagnetic radiation are independent degrees of freedom, and since arbitrarily high frequencies are available, the origin of the conflict now becomes evident: there will be a finite energy to spread over an infinite number of degrees of freedom, so the energy in any one, or in any set of finite frequencies, will be zero; all the energy will be spread infinitesimally thinly over the whole range of frequencies. This is in clear violation of observation. The problem was made explicit by precise calculations of the frequency distribution of radiation at a given temperature in an almost enclosed cavity (so-called "black-body" radiation). The prediction was fine for low frequency, but went wildly wrong at high frequency.

In 1901 Max Planck suggested that this problem would be solved if transfers of energy could only take place in discrete amounts, or "quanta", with magnitude proportional to the frequency:

$$E = h\nu, \qquad\qquad 4.6$$

where ν is the frequency and h is a new fundamental constant of nature called *Planck's constant*. The effect of this was to inhibit transfers of energy into the high frequency modes, and so produce agreement with experiment. The required value of h is about 6.6×10^{-34} kg m^{2}s^{-1}, which, on the scale of *macroscopic* physics, i.e. ordinary experience, is very tiny. (The units used here are the standard units designed to be appropriate for our normal experience, i.e. kilogrammes, metres and seconds.) The smallness of Planck's constant is the reason why quantum effects are not familiar in the everyday world, and is surely at least part of the reason why we find the theory so hard to understand.

Planck's suggestion, although fitting the facts, was very *ad hoc*, and was also in violation of the predictions of Maxwell's equations, so it was regarded as a rather crude substitute for some sort of proper theory. Its significance grew considerably when, in 1905, Einstein showed that the *photo-electric effect* could be explained if it was assumed that light travelled in small packets, which we now call *photons*, each carrying an energy given by Planck's formula above.

The photo-electric effect is the emission of electrons by metals when electromagnetic radiation (e.g. light) falls on them. (The effect is used in all light-meters and most cameras to measure the intensity of the light.) There were several features of the experiments which were impossible to understand on the accepted wave theory of light. First, with a light intensity roughly uniform over a certain area, it was possible to calculate the rate at which energy would fall on any particular atom, and hence find how long it would take for an atom to receive enough energy for an electron to be knocked out. Because atoms are so small this time is considerable, and it would therefore be expected that there would be a delay before any electrons were ejected. Amazingly, no such delay is seen. The electrons appear immediately, when it would seem to be impossible (by many orders of magnitude) for an atom to have received sufficient energy. Einstein realised that this result required that the energy in light was not uniformly distributed, but instead travelled along particular paths (see figure 4.7). In other words the experiment suggested that light consisted of particles! Confirmation of this came from the fact that the energy of the emitted electrons did not depend on the intensity of the light beam but only on its frequency. Explicitly, it was found that the electron energy was equal to the Planck expression $h\nu$, less the energy required to remove an electron from the atom. This is exactly what we would expect if an atom is hit by a particle with energy $h\nu$.

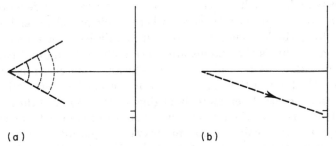

(a)   (b)

Figure 4.7. The photo-electric effect. In (a) the light is emitted uniformly in all directions, so it takes time for a sufficient amount to fall on one atom to make it decay. In (b) the target emits photons which travel in one direction and will hit a particular atom, so decays will be observed immediately.

We have reached the particle/wave dilemma, which is one way of expressing the central problem of the micro-world. There is one set of experiments which, incontrovertibly, shows that light interferes and is therefore a wave process, as predicted by Maxwell's equations. There is another which, equally convincingly, demonstrates that light is a beam of particles. In spite of the remarks by Polkinghorne (1984, pp.7, 8), this problem has not been solved. We shall return to it in chapter 10.

Here we shall continue with the story of how quantum phenomena revolutionised physics. The next big step occurred as a consequence of experiments by Geiger on the scattering of beams of charged particles (emitted from radioactive substances) by atoms.

In 1911 Rutherford showed that these experiments implied that the "currant-bun" picture of an atom, mentioned in section 4.1, was wrong and that, instead, atoms consist of a very small (even on the atomic scale) "nucleus", which is positively charged and which contains most of the mass of the atom, surrounded by enough electrons to cancel the nucleus charge. In fact the radius of the electron distribution, which gives the size of the atom, is about $1 \dots 10 \times 10^{-10}$ m, and that of the nucleus is about a factor 10^{-5} smaller.

The resulting picture of an atom was very familiar to classical physicists, being remarkably like a planetary system, with the nucleus playing the role of the sun and the electrons that of the planets. The force in this case is the electric force between charged objects and, as we have seen, this has the same spatial dependence as the force of gravity. However, whereas this model works beautifully for a single planetary system, it is a disaster for atoms. One reason for this is that atoms are continually colliding and it is inevitable that, as a result of these collisions, the electrons would be knocked out of their orbits and would eventually just fall down into the nucleus. Even more damaging for this model is that as the electrons move in their orbits they would radiate electromagnetic energy at a rate, calculated from Maxwell's equations, which would again force the electrons to fall rapidly down into the nucleus.

We might guess how this problem is solved if we note that its solution requires there to be a *lowest* energy state in the atom, so that when an electron is in the orbit corresponding to that state it cannot lose any more energy. Then we use the Planck relation which allows us to relate this lowest energy to a lowest frequency. This is a very familiar concept; a violin string, a drum etc, all have lowest (fundamental) frequencies. The reason, for example in a violin string, is that the ends are fixed so a wave has to fit exactly into the length of the string (see figure 4.8). This means that there is a maximum wavelength and therefore, since the frequency is inversely proportional to the wavelength (equation 4.1), a minimum frequency. For an electron in an orbit there are no fixed end points, but an analogous condition now arises from the fact that, as we go around the orbit once, we get back to the same point of space, hence again the wavelength must be some integral fraction of the circumference (see figure 4.9). An elementary calculation showing how this gives rise to the energy states of the hydrogen atom is

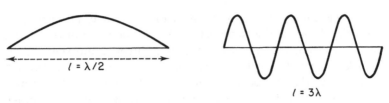

Figure 4.8. Showing how waves on a violin string fit exactly into the length, giving the result $l = k\lambda/2$, where k is an integer.

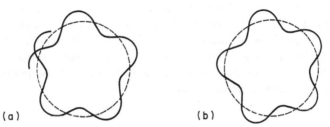

Figure 4.9. Showing how a similar picture to figure 4.8 explains the hydrogen states. (a) is not allowed, but (b) is.

given in Squires (1986, p.147).

Here of course we have extended the wave/particle duality property from photons (which classical theory regarded as waves) to electrons (which were regarded as particles). Indeed, as we shall discuss further in chapter 10, all particles are described by a *wavefunction*, which is a continuous function of position and time. This function satisfies a very simple differential equation, known as the Schrödinger equation after E Schrödinger who first suggested it in 1926. This equation is the quantum theory version of Newton's laws of motion. It applies to all physical systems for which it is possible to neglect relativistic effects (roughly speaking this means that the energy is sufficiently low, but it also excludes massless particles like photons since these are always relativistic). In particular, the Schrödinger equation explains all the properties of atoms and molecules, and hence, at least in principle, all the properties of material bodies. Its first major success, which ensured its rapid acceptance, was the accurate prediction of the spectral lines of hydrogen (recall remarks at the end of section 4.1). It should be noted that there are no free parameters involved in all this agreement between theory and experiment, i.e. there are no numbers which can be adjusted to make the theory fit the data, it either worked or it did not. In the event it did! The only numbers that enter the theory in any significant way are the value of the charge on the electron, which can be measured directly, and the value of Planck's constant, which is determined by one experimental result.

Quantum theory opened up a new field of physics. It initiated the study of atomic and molecular physics, of metals and super-conductors, of nuclei and elementary particles. For further details of the relevant calculations we refer to any of the excellent text-books on quantum theory and its applications. More consideration of what it actually means, of the revolution it has caused in our knowledge of the physical world, and of its significance for the topic of this book will be given in chapter 10. In the next section we shall give a brief account of how our understanding of the microscopic world has gone beyond atoms.

4.4   The standard model

Following Rutherford's discovery of the atomic nucleus it was rapidly realised that all nuclei were made from two types of parti-cle, the proton and the neutron. These have approximately equal mass but only the former carries electric charge, of value equal in magnitude to that of the electron but of opposite sign (convention-ally positive for the proton and negative for the electron).

The most important feature of a nucleus is its electric charge, i.e. the number of protons it contains, because this determines the number of electrons required to make the atom neutral and hence the "chemical" properties of the atom. These are the properties that determine the type of substance it forms, e.g. whether the element is a gas, metal, etc, and how it interacts with other atoms to form molecules. The number of neutrons has a negligible effect on these properties and in fact a typical element can have nuclei with varying numbers of neutrons. These are called *isotopes*. For example, the simplest element hydrogen, for which the nucleus has charge one, occurs as one of three isotopes. In almost all naturally occurring hydrogen the nucleus is just a single proton. However, there is a small proportion in which the nucleus also contains a neutron, and an even smaller proportion in which it contains two neutrons (the elements are then referred to as deuterium and tri-tium respectively). Similarly, most helium nuclei are helium-four, containing two protons and two neutrons, but the isotopes helium-three, with one neutron, and helium-five, with three, also exist. As the number of protons increases, the number of neutrons increases slightly faster, so, for example, the iron nucleus has 26 protons and, most commonly, 30 neutrons, though iron nuclei with 28, 29, 31, 32 neutrons also exist. The highest number of protons in any naturally occurring element is 92, in uranium. These are accompanied by 142, 143 or 146 neutrons, the latter being the most common isotope.

The properties of nuclei were the principle topic of fundamental physics during the 1950's. Although nuclear physicists were able to explain these properties by using the Schrödinger equation of quantum theory, the forces between the protons and neutrons that were required were quite complicated and had to be input to the theory, i.e. they were not derived from anything else.

The physics of "elementary particles", as distinct from nuclear physics, grew with the realisation that in addition to the proton, neutron and electron, which are the constituents of matter, many other particles exist and can be created in the laboratory. The number of such particles, and the pattern of relations between them, increasingly came to suggest that most of them were not really "elementary", but instead were composite objects, just as a nucleus is a composite of protons and neutrons. In 1962 Gell-Mann and, independently, Zweig introduced *quarks* as the elementary particles from which all other strongly interacting particles are made. The idea had some immediate successes (it was indeed designed to fit the pattern of charges, spins, etc of the known particles), but there were many problems with it, and for a long time the "quark model" was not taken very seriously. Indeed, in the 1960's, the fashion was against there being any *elementary* particles at all. "Nuclear democracy" and the "bootstrap" were the current trend. Here, all particles were considered equal; they were all bound states of each other, and it was hoped that their masses and other properties would be determined uniquely by certain self-consistency conditions.

Two things brought about a change in this attitude. The first was a set of experiments at Stanford which, like those on the nucleus more than 50 years earlier, showed that protons really did behave as though they contained small centres of charge, which could naturally be identified as quarks. The other was a series of theoretical developments on, so-called, *gauge theories*. These theories are modelled very closely on the successful theory of quantum electrodynamics (QED), and it seems likely that they are the unique form of consistent quantum theories of particles and their interactions. The particular version that describes the strong interactions of quarks is called *quantum chromodynamics* (QCD) because the forces between the quarks are associated with a special type of "charge", which is universally called "colour", though it has nothing whatever to do with the usual meaning of this word. Indeed this colour property of quarks had been introduced earlier, as a device to solve a certain technical problem associated with the wavefunction of the quarks, and it was only later realised that, by combining it with the gauge theory idea, it solved all the problems of the quark model, and also provided the source of the previously mysterious nuclear forces. The particle that is associated with the interaction, in the way that exchange of photons mediates electromagnetic interactions, is called a *gluon*. One important difference between QCD and QED is that in the former the inter-quark interaction *increases* with separation. This explains why quarks and gluons are never directly observed; they are confined with other particles and cannot ever escape. Indeed the only particles that are seen are "colourless", like the proton and the neutron which are made of three quarks. These should be compared with atoms which are chargeless, i.e. have zero charge. Note that this fact does not prevent their being residual interactions between the colourless states (these are analogous to the van der Waals' forces between neutral atoms).

The other ingredient of the standard model is the Salam–Weinberg theory, which is also a gauge theory, and which provides a unification of electromagnetism with the so-called *weak interaction*. This theory is associated with a "broken symmetry", which has the effect that the particles that mediate the weak interaction, called the W and Z bosons have, in contrast to gluons and photons, non-zero mass. In fact the theory required particular values for the masses in order to agree with the known weak interaction experiments, and it was therefore a great triumph for both theoretical and experimental physics when these two particles, with exactly the predicted masses, were discovered at the European Laboratory (CERN) in Geneva during 1983.

Tables 4.2 and 4.3 provide a summary of the particles of the standard model. As indicated, the top quark and the Higgs boson have not (yet) been seen. They are expected for theoretical reasons. The model has brought fundamental physics to a very curious position. As far as we are able to tell (that is, to the degree that we believe in the validity of the approximations which are crucial in any calculations) the model explains *all* present experimental results. For several years now it has been customary at major conferences on particle physics to have reviews of "beyond the standard model" results. These are things which just might indicate failures of the standard model, but which usually turn out to be either effects of experimental errors or of unexpected corrections to the calculations.

Table 4.2. The spin-half particles of the standard model. Numbers refer to approximate masses in units of the proton mass. Neutrino masses are very small, perhaps zero.

type	charge	1st family	2nd family	3rd family
quarks	2/3	up (.005)	charmed (1.2)	top(>90)(?)
(R,B,G)	1/3	down (.01)	strange (.15)	bottom (4.7)
leptons	−1	electron (.0005)	muon (.11)	tau (1.8)
	0	e-neutrino	μ-neutrino	τ-neutrino

Table 4.3. The other particles of the standard model.

name	charge	mass	spin	associated force
photon	0	0	1	electromagnetic (QED)
gluons	0	0 (?)	1	strong, nuclear (QCD)
$\mathbf{W^+, W^-}$	±1	80	1	weak
Z	0	90	1	weak
graviton(?)	0	0	2	gravity

Higgs boson, with spin zero, associated with symmetry breaking (?)

In spite of this agreement there is no great satisfaction with the standard model as being anything like the final truth. There are features of it that are aesthetically unsatisfactory, it seems to contain too many free, unexplained, parameters and, more seriously, it does not include a proper quantum theory of gravity. Many attempts are being made to find convincing theories that incorporate the standard model as some sort of "low energy" approximation, and that also solve the gravity problem. It is hoped that such theories might also explain the values of some, at least, of the free parameters. The fact that people are dissatisfied with the standard model is in many ways a tribute to its success. We have, in the last 20 years, made such enormous progress, that we have become perhaps unrealistically ambitious! A prescription to calculate numbers is not enough; we would like to see why the ingredients of the physical world are what they are, preferably to see that they could not have been different.

In this connection it is worth mentioning that it is sometimes claimed that there exists a "scientific establishment", which is so attached to a particular set of assumptions that it actively opposes new ideas, and is unwilling to accept contrary evidence. Although there might in some respects be a degree of truth in this claim, it is not generally true. Indeed, with regard to the standard model, exactly the opposite effect is evident. It is universally recognised that the subject urgently needs reliable evidence for something that is genuinely *new*. As I write this, LEP, the new high energy accelerator at CERN, is just beginning to operate. It will enable us to investigate ranges of energy higher, and of distance smaller, than any we have yet experienced. If it does not reveal something that goes beyond the standard model, we will be able to congratulate ourselves on very successful predictions, but we will be disappointed!

4.5   The standard model of cosmology

In the previous section we have described briefly the things of which we believe the world is made. Here we turn to a different set of problems: how did the world come to be as it is; why are its constituents arranged in the way that they are; how is it that the quarks and leptons are in particular elements, in particular proportions; why are there stars and galaxies; why is the universe so big; why does it contain heavy elements; why, indeed, does it contain life?

A crude way of guessing how we go about trying to answer at least some of these questions comes from the simple observation that gravity is a force that is always trying to pull things together. Why is it then that everything in the universe is not falling (or has not already fallen) into some superdense state? Of course we know how the planets of a solar system resist the force of gravity: they are rotating sufficiently fast so that gravity is neutralised by the centrifugal force (compare the fact that an object whirled around on a string does not fall to the centre of its path even though the string is under tension and is pulling it inwards). The universe as a whole, however, is not rotating, so this is not the complete solution.

The answer lies in the fact that all parts of the universe were once moving apart at high velocity, and the effect of gravity over the course of time has simply been to reduce these velocities, but not (yet—see below) to reverse them. An obvious implication of this picture is that if we go back in time we reach a point where all things come together at an infinite-density "singularity". This is the *big bang*, the start of the universe, the point where time and space began, the moment of creation.

The above elementary arguments can be justified in the framework of *General Relativity*, Einstein's elegant geometrical theory of gravity. However, around 50 years ago the idea that the universe started at a particular time was not considered scientifically respectable. Indeed, Einstein was so alarmed by the fact that the simple solutions of the equations of general relativity led to this idea of a universe that was expanding from some initial starting point, that he inserted another term, proportional to the so-called *cosmological constant*, which was permitted by the principle of relativity, and which had the effect of cancelling the overall gravitational pull, and so allowing static solutions of the equations.

That our prejudices have changed so drastically is due to the remarkable discovery of Hubble in 1929, that the universe is indeed expanding. The measurements depend on the fact that light from an object that is moving relative to the observer has a different frequency, the shift being proportional to the relative velocity. (It is not hard to see why this should be so; the waves get spread out or pushed together according to the direction of the velocity.)

Using this means of measuring velocities, Hubble discovered the law that bears his name. This tells us that, exactly as predicted by relativity, distant galaxies are moving away from us with a velocity (V) that is proportional to their distance (R), i.e.

$$V = H \times R, \qquad\qquad 4.7$$

where H is Hubble's constant. We can understand what this means if we imagine a balloon being inflated, and suppose that marks on the surface of the balloon represent galaxies. It is easy to see that the relative velocities all obey Hubble's law. Note that in this picture the two-dimensional surface of the balloon is intended to be analogous to a *two-dimensional* universe. Our three-dimensional universe would in a similar way be related to the three-dimensional surface of a balloon in four dimensions (which we cannot picture). The expansion of the balloon gives the observed expansion of the universe. The picture illustrates one very important point, namely that there is no special "centre". The fact that we observe all galaxies to be receding from us does not mean that we are at a special point in space; rather, observers at any point in the galaxy (surface of the balloon) would see the same thing.

Two questions related to the magnitude of H now arise. First, by running time backwards we can estimate when all the universe had collapsed, i.e. how long has elapsed since the big bang, or, simply, how old the universe is. Roughly, this age is given by one divided by H, i.e. H^{-1}. Actually, it is somewhat less than this because the rate of expansion has been slowing down through the effects of gravity noted above. With an uncertainty of about a factor of two, largely due to the difficulty in estimating the distances to be put into Hubble's law, we find that the universe is 30 billion (or 3×10^{10}) years old. This is an interesting result because it is of the same order of magnitude as the age of the earth calculated by geologists. Considering the magnitude of the numbers involved, and the fact that the two estimates depend on totally different measurements and assumptions, the agreement is remarkable. Of course it is important that the age of the universe is always greater than that of the earth, otherwise we would have a real problem!

The other question is whether the effect of gravity will ever slow the expansion down sufficiently to reverse it, and so bring the universe back to a *big crunch*. Clearly there is a critical initial velocity for this to happen (compare the escape velocity of an object fired from the surface of the earth). Because of the above noted uncertainty in the value of H, and also considerable uncertainties in the average density of the universe, the answer to this question is not known. Certainly the density of the universe is within an order of magnitude, i.e. a factor of about ten, of the critical density.

As we trace the universe back in time towards the big bang it becomes very hot and very dense; the conditions are increasingly far different to anything we have ever experienced. Nevertheless, because we have a good theoretical understanding of the nature of matter, we can try to make realistic calculations. Such calculations have produced two remarkable predictions which are in agreement with observation. The first is that hydrogen is about three times more abundant in the universe than the second most abundant element which is helium. Apart from some minor corrections which occurred much later, this ratio was determined when the universe was about three minutes old. An interesting feature of this calculation is that it would give the wrong result if there were many more light neutrinos than the three standard ones given in table 4.2. Laboratory experiments are now reaching the same conclusion; recent work at LEP (see end of the last section) has just confirmed that there are only three light neutrino species.

The second prediction concerns a relic of the big bang that was formed when the universe was about 300 000 years old. At that stage the universe had cooled to a sufficient degree for electrons and protons to combine to form hydrogen. The photons that were also about at the time did not have enough energy to interact with this hydrogen, and hence they became able to travel freely through the universe. They are still with us, and we observe them as the so-called *micro-wave background*. Their presence was predicted before they were actually observed.

Both the above examples concern the behaviour of the universe before "physics", before life, before planets, or even galaxies were created. It is amazing that we get the right answers! There is one further very speculative calculation involving processes that may have been happening when the universe was a tiny fraction of a second old. Then it was so hot that particles were moving with energies that are way beyond even the most expensive particle accelerators (now or in the future). One of these processes might have been responsible for the present ratio of protons to photons in the universe, indeed for there to be any protons at all. The calculations are uncertain in detail, and depend on some beyond-the-standard-model ideas, but at least they seem to work.

Not everything is successful, however, and there are certainly some problems with the big bang cosmological model in its simplest form. These are, loosely speaking, associated with the fact that it sometimes appears "too good to be true". For instance, we have seen that the universe is a sort of balance between an initial flying apart, and a tendency to collapse. Now we can make an estimate of the lifetime of such a universe, i.e. the time from its creation to when it either becomes so spread out to be uninteresting or collapses down to a point. The basis of this estimate is similar to that which I would use to tell you that if you take a straight pointed stick and hold it as near to vertical as you can, with the point on the ground, and then let go, it will take a time of the order of a second to fall. Now I know that if you held the stick initially in a position sufficiently close to the vertical, then it could take an arbitrarily long time to fall, but, to make any significant difference to my estimate, the accuracy would have to be so fantastic that if you told me that it had taken an hour to fall then I would not believe you. (To make this example realistic you would need to work in a vacuum where there are no draughts.) When we make a similar estimate of the likely age of the universe then we arrive at an answer which is a tiny fraction (10^{-43}) of a second. How then did it come to be so well balanced at the start that it has lasted for ten billion years? (For those who wish to know, the time is calculated from the only parameters that seem to be available, namely the velocity of light, Planck's constant and Newton's gravitational constant, G. From these we form the quantity hG/c^5, which has the dimension of time, and yields the above estimate.)

Another question concerns the micro-wave background mentioned above. We see this coming from different directions in the sky, but from all directions it appears to be the same, i.e. to have the same intensity and distribution of frequencies. Why this should be, when the photons in different directions come from very different parts of the universe, is a great mystery. We might be tempted to think that the answer lies in the fact that, because of the expansion, these distant regions were once much closer. However this does not work. The reason is that signals can only travel at a maximum velocity (the velocity of light) so it takes a certain time to get different parts of the universe to be in agreement. As we go back in time, the parts may get closer, but the time available gets less, and at an effectively much faster rate. This means that we can now "see" distant parts of the universe that in the early stages of the universe could not, ever, have been in causal contact. There is no physical mechanism that could explain why they are the same.

Partial solutions to the above problems are possible by using the *inflationary universe* idea. It is postulated (and various plausible mechanisms exist) that at some early stage in its development the universe became trapped in a phase of extremely rapid expansion. The whole of the presently observable universe was, prior to this expansion, inside one small bubble. Ideas such as this, though seemingly very fanciful, are capable of being discussed, and even used to make testable predictions.

4.6   Unanswered questions

In this chapter we have seen something of the progess that has been made this century in understanding the constituents and structure of the physical universe. We really do understand a great deal, and it is not too much of an exaggeration to say that some sort of an explanation can be given for all observed phenomena. Surely some of these explanations will turn out to be wrong, and it must certainly be hoped that there are new things to be seen that will surprise us. The general picture, however, is, I believe, likely to be correct, in so far as it goes.

What then are the remaining problems? These can usefully be classified into several groups.

First there are what might be called technical problems. These occur when we have a theory which we believe is correct, but which is too complicated to permit us to calculate with sufficient relia-bility to compare with data. A good example here is QCD, the theory of strong interactions, which we believe, but cannot prove, explains the whole of strong interaction physics. Of course it may be that, in some cases, if we could calculate, we would find that the theories failed to work. These problems would then come into the next class.

These are the places where we really do not have a satisfactory theory. Without question the most serious example here is that we do not have a proper quantum theory of gravity. Gravity is a very weak force, so for most purposes we can use what are called "lowest order" calculations, and obtain perfectly adequate results. This is fortunate because the higher order "corrections" often turn out to be infinite! Apart from the fact that this is a very unsatisfactory situation, we have no means of understanding what happens when both quantum and gravity effects are important, e.g. very close to the time of the big bang. It is very likely that the "singular", infinite density, point associated with the big bang does not really happen in a proper quantum theory, but this issue is far from being properly understood.

The next class of problems are more *aesthetic*. Certain aspects of the standard model are very unnatural, so there really ought to be a deeper explanation. *Why* is the gauge group of the standard model $SU(3) \times SU(2)_L \times U(1)$, rather than something else? (Readers who do not know what these symbols mean need not worry. They probably have no less chance of answering the question than anyone else at present.) Why are there three, or whatever, generations? Why do some of the interactions respect parity (look the same in a mirror-image world), whilst others do not?

Then there is the recently much discussed fact that the existence of a world that can contain people seems to require a lot of apparent "accidents" in the values of the parameters of nature. It looks as though these have been "fine-tuned", often with amazing accuracy, just so that life can exist. We again must distinguish between two types of requirement (cf. section 2.3); on the one hand there are things that are essential for living beings to be *possible*, and on the other hand there are a different set of requirements which are important for life actually to *occur* in the world as it is. As a simple example of the former, it seems as though life, or indeed any interesting structure, requires the existence of chemistry, i.e. materials containing heavy nuclei. The possibility of such nuclei depends critically on the strength of the strong interaction, and its relation to the electromagnetic force. One example of the conditions being right for life actually to occur has already been described in the last section; namely, the original expansion rate of the universe has to be correct for the universe to last for a sufficiently long time. Another, this time associated with the fundamental constants of nature rather than the initial conditions of our universe, is that the production of heavy elements in stars depended in a very precise and subtle way on particular energy levels in carbon and oxygen atoms. In some sense an even more remarkable example of fine tuning is the fact that the cosmological constant, again mentioned in the last section, is known to be less than 10^{-120} when measured in the only natural units that are available. This might well be described as the most accurately known number. We have no real idea why it is so small.

The problems in the last paragraph are often discussed in terms of the *Anthropic Principle*, a name given to a variety of related ideas. One aspect is that we should not be surprised that the parameters in the world are just right to contain us, because, if they were not, then we would not be here to discuss the issue. The fact that we are here forces the parameters to be right. It is easy to see how this might explain the fact that the local conditions on the earth are suitable—we clearly live in a place where the environment is right—or why the universe is as old and as large as it is—this comes from the necessity of stars to have formed and collapsed in order to spread heavy nuclei around. Here we are are using what is called the "weak" form of the anthropic principle. But this does not help us to explain why there has to be *any* time or place that is suitable for our existence. For this we either need to assume that there are many different universes, with different values of the parameters, and maybe different forces, particles, etc, so that it is not unreasonable to believe that there is one in which we can exist, or we must assume the *strong* form of the anthropic principle, in which our existence is somehow necessary for there to be a universe at all. Of course if it was *designed* with us in mind then this would not be so much of a problem! Some of the parameters, however, like the cosmological constant, seem to be even more accurately fixed than our existence requires. We refer to Davies (1982) and to Barrow and Tipler (1985) for further discussion of this fascinating topic.

The last paragraph is related to the question of how much we should expect of a complete theory of physics. Many of the recently postulated TOE's seem as though they ought to fix the actual values of all the parameters, i.e. to say that only one "physics" is possible. But surely this cannot be so, because it would then be inconceivable that such a physics should contain us! (I am not sure what probability really means in this case but there seems to be little doubt that "most" universes would be very uninteresting! See, for example, the discussion of Bartholomew, 1988.)

There is another question about the physical universe, which I am not sure is a sensible question, but which I sometimes think about. We have, for example, said that Newton calculated the orbits of the planets from his law of gravity. But how do the planets themselves calculate their orbits? How does a planet know where to go next? Putting this another way, we have not yet found it possible to calculate the mass of a proton from QCD. But millions of three-quark systems do it instantaneously. How? One sort of answer to this type of question is to say that the things of physics do not actually *do* anything, they just *are*. Newton's famous first law of motion is of this type: "a body continues in its state of rest or uniform motion ...". But it cannot end there. Newton had to add "unless acted upon by a force ...". It is not easy to see how *constancy* rather than *change* can be the fundamental property of the universe, so how do objects "know" in what way they must change?

We close this list of problems by reminding readers that we have still to discuss the deep problems raised by the quantum nature of physics, and of course we have not said anything in this chapter about where, or how, or if, there can be any place for the main subject of this book, conscious mind, in the physical world.

Chapter 5

Philosophical background

5.1 "Words, words, mere words"
(Shakespeare in Troillus and Cressida)

The observation of the basic properties of human consciousness requires no equipment, no knowledge of scientific theory, no particular training or technical ability; even the most primitive man is aware that he is aware. Thus it is not surprising that the study of mind has a long history. However, whereas the development of scientific techniques has increased our ability to make detailed observations of the outside world, with the consequent increase in our understanding as discussed in the previous chapter, there has been no comparative growth of observational facts in the field of consciousness. The study of mind has remained very much "in the mind". Hence the subject has been dominated by opinions and fashions, and, in spite of the millions of words that have been written, there has been very little progress. The contrast with the previous chapter is striking. Our failure to understand schizophrenia, which appears to be associated with aspects of the brain very close to consciousness (Janes, 1976, Claridge, 1987), is perhaps the most tragic aspect of this lack of progress.

In this chapter we endeavour to review the more important ideas that have emerged over the years. Inevitably this will involve a degree of oversimplification, and our discussion will be somewhat idiosyncratic. The aim is to give a physicist's understanding of what some of the basic questions are about. We shall proceed by introducing the words that are frequently used in the debates about consciousness. These words describe general classes of ideas, so their definitions are very loose; they convey an "ambience" rather than a precise position, a fact which does not always prevent their being used in dogmatic statements! I have tried to give a balanced and fair account of the various views but, inevitably, my own prejudices will colour what is written.

Readers who require more detailed (and authoritative) treatments of many of the topics mentioned briefly in this chapter might consult, for example, *The Oxford Companion to the Mind* (Gregory, 1987), which is referred to in here as OCM, or *An Encyclopedia of Philosophy* (Parkinson, 1988). A useful book, written for "newcomers to the philosophy of mind", is *Matter and Consciousness* (Churchland, 1984). A modern, and very careful, though admittedly partisan, account of the subject is given in *A Theory of Determinism* (Honderich, 1988).

We begin with the view of reality which most emphasises the significance of conscious mind:

5.2   Idealism

The basis of idealism is the simple observation that all knowledge comes from sensations in the conscious mind. Thus, since everything I know, I know through my mind, it follows that, in some sense, my mind is the only certain reality. In the words of Flew (1984) idealism is *A name given to a group of philosophical theories, that have in common the view that what would normally be called "the external world" is somehow created by the mind.* Various versions of idealism differ in the degree of reality they admit for material objects, i.e. for the external world; possibilities range from the form in which we recognise them being a product of the mind, through the external objects being created by the mind, to an extreme view which would say that mind is all that there is. In philosophies of this type consciousness is the true reality, the world "out there" has no intrinsic reality, no existence apart from the fact that it exists in my conscious mind.

  The Encyclopedia of Philosophy (Edwards, 1967) defines idealism more generally, and more emotively: it *is the view that mind and spiritual values are fundamental in the world.* Although this definition is probably closer to the normal use of the word "idealism", it seems to be too general to convey the full significance. Indeed even a realist (see below) would not *necessarily* argue with the claim that spiritual values are fundamental. We shall therefore use the term in the sense of Flew's definition given above.

The general ethos of idealism is opposed to physics, at least it is opposed to physics seen, as I see it, as an attempt to understand the external world. Such an attempt must fail, because there is no external world! Rather, to an idealist, the true explanations of my apparent experiences of an outside world are to be found in an understanding of my conscious mind.

There are other philosophical schemes which, at least for the purpose of our discussion, we can regard as belonging within idealism, in particular because they have a similar attitude to physics. For example, **instrumentalism** is based on the premise that scientific theories are not explanations, but are simply tools for calculating a particular set of experimental results; they are useful in so far as they give the correct answers, but *there is no question as to the truth or falsity of these theories—they can be neither true nor false* (Flew, 1984, p.125).

Another related set of ideas goes under the name of **positivism**, in which it is claimed that observed phenomena are all that we can properly speak about, and that other things, e.g. underlying truth, explanations etc, which are not subject to experimental test are meaningless. At first sight positivism seems to support the scientific method and indeed positivists certainly regarded themselves as pro-science, contrasting the rigour and certainty of scientific investigations, based on observations, with what they regarded as the essentially useless concepts of religion or metaphysics. Ironically, the idealism which is implicit in positivism means that in practice it is not favourable to scientific investigation and progress. An illuminating example of this (discussed by Nyhoff, 1988) is the early opposition to the "kinetic theory" of thermodynamics, which was largely motivated by philosophical rather than strictly physical considerations. To a positivist, the rules of thermodynamics gave agreement with data, so who needed an "explanation"? Ernst Mach, a convinced positivist, who once wrote that *the world consists only of our sensations*, was an extremely influential physicist opposed to the kinetic theory as an explanation of thermodynamics: *the view that physical phenomena can be reduced to processes*

of motion and equilibrium of molecules is so universally spread that
... one can only let people know that ones convictions are opposed
to it. (We note that this remark can also be seen as an invalid (!)
criticism of reductionism.)

In fairness to Mach, it should be noted that he also argued
against a simple "mechanistic" explanation of everything, e.g. of
electromagnetism, and here he was rather in advance of his time
and had a beneficial influence on the work of Einstein. Nevertheless it is quite hard for us to realise that the quotation in the above
paragraph was written only just over a hundred years ago. It is
even harder to contemplate what the world would be like today
had the negative blanket of idealistic (or positivist, or instrumen-
talist) thinking prevented scientists from asking for explanations,
and seeking understanding of the observed world. This is probably
the most useful argument against all such philosophies; they discourage endeavour to understand the sensations of the conscious
mind. Even if few idealists would quarrel with the statement that
I felt a pain in my head because a stone hit it, they would be reluctant to seek similar explanations for the more subtle manifestations
of external reality.

I have given elsewhere my own reasons for rejecting idealism
(Squires, 1986) and I shall not repeat the arguments here. I believe
that it is logically unassailable but, in practice, foolish and sterile.
There are few things stated with any conviction in this book, but
this is one of them! Modern physics has taught us that "reality"
is an illusive concept, but this will only encourage us in our belief
that it is a worthy aim to search for it! Here we are "realists" and
hence adopt the philosophy of

5.3   Realism

This view accepts that our conscious mind receives experiences
from a real external world. The images we obtain, through various
observational methods, involving for example eyes, ears, telescopes,
etc, are images of a genuine existing reality, whose existence is
not dependent on our being aware of the images. Of course we
ourselves are part of the real world and in the act of observing it
we disturb it, but this is not to say that we create it. (Incidentally
it is sometimes suggested that the disturbance of the world by

observation is a quantum effect. This, of course, is not true, as the disturbance occurs even in classical situations. The extra thing in quantum theory is that in certain cases the degree of disturbance cannot, even in principle, be made arbitrarily small.)

Flew (1984) defines realism as the belief that *physical objects exist independently of being perceived.* Thus a realist is interested not only in questions like: *What do I observe?* but also in *What is?* A realist has no difficulty in accepting that the world actually does contain other things and other people, and that these are made of atoms containing electrons and protons, etc. He believes that planets, stars and galaxies exist, and that history is the story of things that really happened to real people. It is surely true that most scientists are, and always have been, realists. Somehow the whole scientific enterprise loses its essential motivation, the intellectual fun goes away, if there is not a real world to explain.

It must be admitted, however, that quantum theory is a big shock to a realist philosophy. As we have already suggested, and will discuss in more detail later, quantum theory is excellent for telling us what we will *observe*, but runs into serious trouble when it tries to tell us what *is*. This fact probably encouraged the wholesale retreat from realistic philosophy which has been characteristic of much of this century, and which indeed persists. Although probably only a few philosophers would now support the extremes of idealism, many seem to seek a sort of intermediate position (see, for example, von Fraassen, 1980), which is very far from the realistic way of thinking of most scientists, and the statement: *realism is well and truly dead* appears in a recent book (Fine, 1986).

In my opinion this is the wrong reaction to the enigma of quantum phenomena. Rather than running away from the problems by abandoning the search for realism, we should recognise that we will have to be much cleverer. The real world is not as simple as we once thought. We have learned that "naive realism" (a term used by d'Espagnat, 1983) is certainly not correct. Indeed, it might even be that by following this method we discover that there are aspects of our experience which, though we might want to see them as part of an external reality, force us to believe that they are in fact created by the human mind.

The situation here is rather like that with reductionism, discussed in section 2.3. Realism is an *aim*; we try to understand a real world, existing independent of our conscious mind and explaining the sensations it receives. If at some point we find that we cannot do this (and we shall not give up without a struggle, and will be prepared to think very carefully about what we mean by "physical objects" in Flew's definition given above), then we really will have made an exciting discovery!

A spirited and provocative defence of realism, *Is realism a dirty word?*, is given by Gardner (1989); more serious readers might prefer that given in the book *Computational Philosophy of Science* (Thagard, 1988). Unfortunately neither of these authors give sufficient attention to the real problems presented by quantum theory. For this one should consult the detailed discussion of d'Espagnat (1989) in *Reality and the Physicist*.

We return now to the central topic of this book and note that, whereas idealism clearly has an important place for conscious mind (there is *nothing* else), realism has nothing specific to say on this point, apart from asserting that there *is* something else. However, this already reveals the crucial issue, namely the relation between my conscious mind and the reality that is external to it. The latter certainly influences my conscious mind, e.g. the temperature in this room has started to drop and I am aware of feeling cold. Apparently there is an influence the other way, e.g. because I felt cold, I closed the window. But what is the nature of this influence? Is it through the laws of physics or does it violate them? Is it in fact only apparent? Is my conscious mind, seen by you as part of your external reality, something beyond the "physical world", or is it just an aspect of physics? If we are inclined to answer yes to this last question, then we are supporting the doctrine known as

5.4 Materialism

The basic statement of materialism is that *matter* is the only reality. Again quoting Flew: *whatever exists is either matter or entirely dependent on matter for its existence.* In fact, this is probably a rather out-dated way of expressing what materialism really means; it belongs to an age when the physical universe was regarded as consisting exclusively of "matter", i.e. solid objects interacting with each other through contact, or "pushing". As we have seen, physics, even classical physics, has now abandoned such a simplistic description of the material world, so a better statement

of materialism would be to replace "matter" in the above definition
by "the laws of physics" (for this reason the word "physicalism"
is sometimes used as an alternative to materialism). The OCM
says: *materialism is best interpreted as the doctrine that the fundamental laws and principles of nature are exhausted by the laws
and principles of physics.* Of course this leaves open what we mean
by the laws of physics, and there is a danger that the statement
might be nothing more than a tautology. It would be more explicit
if we referred to "the laws of physics as could be deduced from the
interactions of quarks and leptons" or even "as defined by the standard model". Materialism is the belief that the TOE's, which we
mentioned in section 2.1, really are theories of **everything**; *there
is nothing else.*

An alternative version of materialism would not deny the possibility that there was something else, other than physics, but it
would assert that whatever this might be it can have no effect upon
the physical world. Popper calls the world of physics *World 1*, and
the two versions of materialism are that *World 1* is either all that
there is, or that it is causally closed, i.e. cannot be affected by
anything outside it.

However difficult it may be to make a precise definition, it is
clear that materialism is in sharp contrast to idealism; it tends to
diminish the significance of "mental" phenomena; whatever these
are, they are not regarded as primary. Also the materialist view
is normally associated with a denial of spiritual or religious values
and concepts, and consequently with a rejection of the idea of God
(see, for example, the article of Atkins, 1987). It is not completely
obvious that there is any logical reason for this, and I note that
John Polkinghorne, an ex-theoretical physicist, now a Church of
England minister, who regularly writes as an effective apologist
for the Christian religion, is to some degree at least a materialist.
He would, I am sure, deny this, but his statement (Polkinghorne,
1988, p.71): *If you take me apart you will find that all you will get
will be matter* ... seems to be one way of expressing the basic materialist doctrine that there *is nothing else.* In the following pages
it is clear that Polkinghorne sees no incompatibility between this
doctrine and a basically religious understanding of reality. Similarly, I am not convinced that the facts that men are *"ends in
themselves"* or that *"we value human lives"* (Popper and Eccles,
1977, pp.3, 4) are incompatible with materialism, or that there

is any obvious contradiction in materialists being *"humanists and fighters for freedom"*.

There is a basic simplicity and economy of ideas in materialism that makes it extremely plausible. It receives strong support from the enormous success in providing explanations that physics, and science in general, have had. Even something as fundamental as *life*, which was once thought to lie beyond our powers of comprehension, would probably be regarded by most contemporary biologists as being just another manifestation of the many forms of matter: *organic life contains no principle not already in non-organic matter* (OCM, p.491). Why therefore should we not be content to say simply **man is a machine**?

The difficulty for materialism comes when we try to discuss the topic of this book: consciousness. Materialism is happy with light of wavelength 5.9×10^{-7} m, but where within it can we find "yellow", which is the way I experience such light. Many words have been written in an attempt to answer questions of this type. Sometimes the answers seem dangerously close to saying the topic is difficult so we will talk about something else, for example pointing out how inadequate are non-materialistic explanations! Even worse of course is to deny the existence of the things of conscious mind. In the article noted above, for example, Atkins suggests that *we accept, at least as a working hypothesis, that the world is barren of purpose.* Presumably there is some "purpose" behind this request, so it contains its own reason for rejecting it!

A more realistic approach, known as **periphalism** or **behaviourism**, asserts that our behaviour is totally determined by materialistic considerations not involving consciousness (i.e. by the laws of physics), and that the existence or otherwise of conscious mind is either totally irrelevant and therefore of no interest, or else can and should be discussed solely in terms of behaviour. Consciousness is therefore merely an **epiphenomenon**, of no more significance to the real world than the steam whistle on a train is to the working of the engine. Such a view, of course, is not much help to us here since our aim is the discussion of consciousness. Also, if it is saying, as would seem to be the case, that the existence of consciousness has no effect on the physical world, then it can be rejected on the basis of the conclusion reached in section 3.4. E Honderich refers to this as the *axiom of the indispensability of the mental* (Honderich, 1987, p.447), and justifies it by pointing out what he refers to as the *futility of contemplating its denial.*

"Behaviourism" in fact originated in the work of a psychologist, J B Watson, and was there seen just as a method of doing psychological research, i.e. trying to analyse and influence behaviour without bothering what the person or animal was thinking. As such it has had, and still maintains, a big influence, although I believe that even here it has been realised that it is often inadequate—properties of conscious mind tend to come back into the discussion.

Certainly as a way of understanding conscious mind, behaviourism now appears to be discredited. Its place, within materialism, has been taken by what is now probably the most popular theory of the relation between mind and matter, and which therefore deserves another section, namely the **centralist** position, more usually known as:

5.5   The mind–brain identity hypothesis

One hot Sunday morning many years ago I attended the service at the Anglican church in Trieste. It was Trinity Sunday and consequently I met, for the first time, the creed of St. Athanasius. This largely consists of repeating, in several different ways, the essentially absurd claim that three "persons" could be three separate individuals and yet be one indivisible being: *For there is one person of the Father, another of the Son: and another of the Holy Ghost. But the Godhead of the Father, and of the Son, and of the Holy Ghost is all one.* We were expected to believe that something could become true, if it was asserted sufficiently often that it *was* true.

Now it is easy to understand the theological problems which met the early Christians and which led them to this strange contradictory doctrine. They were spiritual (at least) descendents of Judaism and if there was one thing on which the Jewish religion was adamant it was on the essential uniqueness, "oneness", of God. The gentile religions were characterised by a multitude of gods, and the prophets of Judaism had had a continuous struggle to prevent such a concept from tainting the purity of their monotheistic faith. But these people had now met, or had heard about, Jesus the prophet of Nazareth, and they had come to regard him as divine. Also, and it is harder to trace the line of thought here except that going from two to three is less of an effort than going from one to two, they found it desirable to regard the apparent power of God in their lives as the effect of another divine person, the Holy Spirit. Thus they found that they had three "divine" persons, i.e. three separate and different Gods. But there was only one God, so the doctrine of the Trinity followed naturally: one God is identical to three Gods!

The mind–brain identity hypothesis seems to me to be a very similar sort of idea, and to have a similar motivation. Roughly speaking it claims that the problem of the relation between the conscious mind and the physical brain is not a real problem because these two things are, it asserts, not two different things which need to be related, but are just the same, single, thing. The fact that they are *not* the same, and that *The problem, indeed, is to find anything in common!* (Gregory, 1981, p.477) is ignored, with a carefree disregard of the obvious, reminiscent of that in the Athanasius creed, where the fact that one is not equal to three is similarly treated.

According to the identity hypothesis conscious thoughts *are* physical brain states; not, we emphasise, *are caused by* or *are associated with* or even *cause*, but actually *are*. Thus, what we might be tempted to think of as two separate things, is actually one thing. Although the idea in its modern form is attributed to Feigl (1967), similar ideas were already expressed much earlier by Romanes (1896): *any change taking place in the mind, and any corresponding change taking place in the brain, are really not two changes but one change.* As an analogy Romanes suggests the seeing and hearing of a violin, for which he asserts ... *we hear musical sound, and at the same time we see a vibration of the strings. Relative to our consciousness, therefore, we have here two sets of changes, which appear to be very different in kind; yet we know that in an absolute sense they are one and the same.* Even within its own terms this analogy seems to be simply wrong; we cannot identify the visual and auditory signals, light and sound are not the same, and indeed it would be quite easy to have one and not the other. What is true in this example is that the two signals have a common origin. Thus this example is not really an analogy for an "identity" theory, but for something closer in spirit to the "nomic connection" ideas of Honderich (1987, 1988), which we discuss in the next section.

In the above references Honderich gives more detailed criticisms of various versions of the identity theory. In particular he claims that such theories are not compatible with the assumption, mentioned above, of the indispensibility of the mental. He concludes (p.451 of Honderich, 1987) that *virtually all contemporary Identity Theories entail the existence of two of something* and that *nothing whatever is gained by saying it is one thing that has certain mental and physical properties, and any theory of mind that comes to no more than this must fail.*

Searle (1984) proposes a solution of the mind–brain problem which, though he explicitly denies is an identity theory, is certainly sufficiently close that it can conveniently be discussed here. He makes two assertions:

> *mental phenomena ... are caused by processes going on in the brain;*
> *mental phenomena just are features of the brain*

In other words minds are caused by brains and are also features of brains.

As an analogy Searle refers to the way solidity is caused by the properties of molecules and their interactions, and is at the same time a property of the molecules. The fact that a given object maintains it shape can be attributed either to particular features of the intermolecular forces, or equally well, simply to the object's solidity. There is really only one explanation, though it can be expressed in different ways.

If we use this analogy then we would suppose that being conscious is a property of a collection of molecules in the same way that being solid is. The question: *What particular physical structures are conscious?*, would be no different in kind to the question: *What particular physical structures are solid?* This is an interesting analogy, but it seems to me to avoid the issues. Being solid clearly is a physical property; it refers to how the molecules move when they are subject to physical forces, and every aspect of it can in principle be calculated from knowledge of the properties of those molecules. However, consciousness is not *clearly* such a property; indeed the most obvious indications are that it is not such a property at all. We cannot calculate whether a given system of molecules will be conscious, and the problem is not anything to do with the difficulty of the calculation, it is rather that we have no idea even how to begin! Using the notions of the identity hypothesis, it is quite reasonable to say that the property of solidity is in some sense "identical" to the property of one molecule not being able to move relative to another; these are genuinely different ways of saying the same thing. It seems much less reasonable to say that the sensation of "red", for example, is identical to any obviously physical property of a set of molecules. Certainly it would appear to be a requirement on those who believe this to suggest what the property is, and why. Pippard (1988) confidently asserts: *What is surely impossible is that a theoretical physicist, given unlimited computing power, should deduce from the laws of physics that a certain complex structure is aware of its own existence.* Is it just our lack of imagination that makes this statement appear to be true? If not, then while we might be materialists by prejudice, we have nothing on which to base our belief. The conclusion of Churchland (1984): *The weight of evidence ... indicates that conscious intelligence is a wholly natural phenomenon. According to a broad and growing consensus among philosophers and scientists, conscious intelligence is the activity of suitably organised matter ...* is too strong. If it is true that there is a "growing consensus", then it is based on difficulty with other ideas, and not on "evidence".

As I have already suggested, the motivation behind the identity hypothesis seems to be very similar to that behind the doctrine of the Trinity. We hold two contradictory beliefs, by asserting that they are not contradictory! We cannot deny the reality of consciousness, so, if we accept that there is also a real world of physical things, we must either say that consciousness *is* one of those physical things, or that it is something else. The latter is as objectionable to materialism as the idea of "many gods" is to monotheistic Judaism. It implies that, apart from, and different to, the objects of physics, there are *other things*, of which conscious mind is one. It is an idea which, like multi-theistic religions, goes back a long way in time, and which is usually referred to as

5.6 Dualism

This is a name given to the class of theories which would answer the question: *Is there anything else?* with an emphatic "yes". Dualism distinguishes two types of substance: the "material", and here it is presumably proper to allow this to include all that is included in materialism, so the "physical" is a better description, and the "mental".

The credit for first making this distinction explicit is universally given to René Descartes, a French mathematician, philospher, scientist, etc. He lived in an age (1596–1650) when it really was possible to be all these things in a serious way! It was a time when science as we know it was just beginning. Observations were being made and, although Newton had not yet shown us how to understand planetary motions, the possibility that *natural phenomena could be scientifically explained* was being realised. Descartes accepted these scientific, materialistic, explanations, and, in particular, was a pioneer in regarding even life, i.e. the phenomenon of organic matter, as being not essentially different to the properties of inorganic matter. However, he realised where this materialistic philosophy was leading, and presumably saw the "threat" noted in section 2.2. Scientific laws might explain the behaviour of material objects, but they surely could not explain people. Hence, for humans, but not animals, he postulated the existence of something new, something outside the operation of physical laws.

Descartes worked on an ambitious endeavour to base all human thought on precise and provable statements. This got him as far as his famous proof of his own existence: *I think; therefore I am.* (Many centuries earlier Augustine had reached a similar conclusion by arguing that only existing things can make an error, so it cannot be an error to assert that I exist.) To proceed, he was forced to introduce some assumptions, which then allowed him to give a "proof" of the existence of God. Further progress depended on assuming the basic "reasonableness" of God. One line of argument from the proof of his own existence led him to the conclusion: *I am a being whose essence is to think and whose being requires no place and depends on no material thing.* Thus a person is essentially a conscious, thinking, substance, unextended and indivisible, in some sort of association, but not dependent upon, a physical body.

All this is the complete opposite of materialism. We have moved from the idea that I am a physical object to the idea that I am independent of any physical object. Again, but of course for different reasons, the idea is very appealing. It is how we most naturally think of ourselves. I speak of *my* leg, *my* heart, *my* body, etc;

none of these things do I naturally regard as being *me*. Dualism allows a clear statement of who is conscious; certainly it is in no doubt about the fact that wholly physical things (machines) cannot be conscious. In one sense it tells us what consciousness is, it *is* the thinking substance, not describable in terms of other things. Also, dualism has the obvious appeal that it is readily compatible with the possibility of *me* surviving the death of *my body*. Indeed this fact has been made the basis of a "proof" of the survival of a person after the death of the physical body (see, for example, Swinburne, 1987).

Nevertheless dualism, sometimes known as Cartesianism, is strongly criticised by many philosophers and at least in its obvious form *substantive dualism, the doctrine that the mind is a separate, non-physical entity, now has ever fewer supporters* (OCM, p.190). Indeed (like reductionism, but by different people) it is sometimes used as a term of abuse; for a particular view to involve dualism is regarded by some as a reason to reject it. On the other hand there are those who maintain (and regret) that dualism still has considerable influence on contemporary philosophy: *Cartesianism has such a grip on philosophical thinking that opponents appear either extravagant or mad* (Bakhurst and Dancy, 1988). For a recent critical review of Dualism we refer to Smith and Jones (1986) and to Churchland (1984); for arguments in its favour to Swinburne (1987).

The OCM also refers to an alternative form of dualism known as "attributive dualism". Here there is assumed to be nothing other than the physical brain, but this is given two sorts of attributes, the obviously physical and the mental. Some attempt is made to draw an analogy between the hardware and the software of a computer. It is not clear to me that all this means anything significant, or that it is properly referred to as a dualistic theory. There is not really any fundamental difference between the software and hardware of a computer. They are both clearly physical things, one built into the machine when it was made, and the other pushed in on a floppy disk.

The reasons behind much of the opposition to dualism are easy to see. It requires there to be something that the laws of physics cannot explain, yet, as we have stressed, the ability of those laws to explain things has steadily grown, and the ground occupied by that which is "outside physics" seems to be continually shrinking.

Why should we believe that human minds are in any way different? Then there is the fact, discussed in section 3.4, that many of the behavioural features that we want to attribute to consciousness in humans are also seen in animals, which, at least to Descartes, are not conscious. If, on the other hand, we extend the idea to allow animals also to possess this "mind" substance, we run into the problem of how far we go; there seems to be a continuous line of behaviour running from plants, through the lower forms of life, and on to humans.

The most serious difficulty with the dualistic idea, however, is certainly the one which perplexed a young Stuart princess, Elizabeth of Palatine. She was the eldest daughter of the "Winter-King" Frederick of Bohemia and his wife, also called Elizabeth, who was a daughter of the Scottish King who became James the First of England. Her father's brief period as King of Bohemia was ended by an Austrian army; an army which, incidentally, included Descartes among its members (a fact which suggests that he was not as intelligent as his reputation implies). In consequence, Elizabeth was in poverty, at least by the standards that were expected for a princess. Since she was also a devout Protestant, at a time when all the available suitors were Catholics, she saw, even as a teenager, that her marriage chances were small. Being, unlike her mother, of a naturally quiet temperament, she consoled herself with the study of philosphy and science, and hence she would be delighted to have the opportunity to talk with the now famous Descartes when he was introduced to the family at the Hague in 1642. There began a friendship, carried on apparently through correspondence, that lasted many years, as Elizabeth tried to understand the ideas about which Descartes was writing.

One problem, in particular, troubled her: she failed to comprehend in what way the thinking soul could possibly influence the body which was not thinking. Although Descartes, doubtless flattered by her attention, was enthusiastic with praise for his young pupil, both for her beauty and for her brains, and tried to satisfy her curiosity, it is doubtful whether he was ever successful with this problem. One particular suggestion that he made was that the source of the influence of mind upon matter lay in the pineal gland. It seems that he chose this site, deep in the centre of the head, because there is only one such gland, whereas most of the features of the brain are in pairs. The suggestion, however, is not such as to encourage contemporary minds to take Descartes seriously!

Elizabeth's problem remains as a basic difficulty with dualism. How can we understand the "connection" between the mental substance and the physical body? It seems obvious that there has to be such a connection. In particular, it is surely reasonable to say that our conscious minds are affected by what happens in the physical world, i.e. by our sensual experiences. We could, in principle, dispute this of course. Just because two things are correlated in time does not *necessarily* imply that one causes the other. For example, when I look out of my window I see trees turning gold with the approaching winter; that is, (1) light of a golden colour falls on my eyelids and (2) I become aware of seeing gold. Fact (2) could be independent of fact (1) in the sense that it may just happen that the two events occur simultaneously. It would of course be necessary that such apparent coincidences are the rule, not the exception; essentially there would have to be two parallel worlds, the mental and the physical, always closely synchronised yet unaffected by each other. I mention this possibility because in various forms it is extensively discussed in the literature, but we can surely dismiss it. Indeed it is easier to accept the extreme of idealism, which dispenses with the unnecessary physical world altogether.

Returning to more reasonable ideas, we accept that the physical world has an effect upon the mental. Although it is perhaps hard to see how this might come about, it does not cause any major difficulties, essentially because we have no laws to describe the behaviour of the mental substance. However, it is natural to suppose that there is also an influence going the other way. We are conscious of the desire to do something and can translate that desire into the particular action. This again is how things *appear* to be. Thus the mental substance can affect the physical. The situation is here very different because we *do* have laws to explain how the physical world behaves, and these laws seem to tell us *everything*. There just is no room for any possible effects that consciousness might have upon that world. In other words, as we noted in section 5.4, the world of physics, i.e. Popper's World 1, is "causally closed"; according to our present understanding, it cannot be altered by anything from outside of itself. In one sense the problem was not seen to be as serious at the time of Descartes as it became later, because the explanatory powers of physics were not known, and it was easier then to accept that the world of physics was not closed. More recent events, however, may have changed things the other way. The simple idea of forces being just the push–pull of contact between material bodies has gone, and the advent of quantum theory has completely changed our views on the degree to which the physical world is closed.

A recent attempt to combine the best features of identity and dualistic theories is due to Honderich (1987, 1988) who introduces the idea of

5.7 Psychoneural pairs

Each of these consists of a *neural* and a *mental* component, in what Honderich refers to as **nomic correlation**, such that one is an essential and necessary partner of the other. A possible metaphor here might be to think of an object which necessarily has both a back and a front. In fact the relation is a *one-to-many* relation because a class of neural events can correspond to the same mental event. It is postulated that these pairs are the causal agents which determine the operation of the brain, and the model therefore gives a natural explanation of the synchronicity between mental and physical events, without requiring that the physical world is non-closed. The model is dualist in that it asserts the existence of the mental *in addition* to the neural aspect of the pair, but this does not carry any of the metaphysical implications of Cartesian dualism. It is also materialistic in its insistence that the mental cannot exist without the material. Indeed, although the mental is distinguished from the neural, there is no requirement that the mental component should be *non-physical*. Further details, including an endeavour to define "physical" as applying to anything with location in space, are given in the references quoted above. To some extent I see here again the difference of emphasis between a physicist and a philosopher; for example, there does not seem to be any attempt to answer the important question of when an apparently physical system changes into being a "pyscho–neural" pair. What are the key physical things that make this happen? Although Honderich clearly believes that the model satisfies the principle of the indispensibilty of the mental, noted above, it is not obvious to me why this is so. Presumably we could calculate how the psychoneural pairs will interact and evolve in time (here we note that Honderich is claiming to justify a deterministic operation of the brain, so no obscure quantum effects should be allowed). Such calculations will make no reference to mental events, and will not tell us anything about when they occur, or in what form. How then can they have an effect?

In general, we can characterise the distinction between the materialist (monist) position of section 5.5 and the dualistic position of this section by saying that in the former *some particular types of physical object* are conscious, whereas in the latter (at least in the traditional form) *no physical object* is conscious, though some might become in a rather obscure way intimately associated with consciousness. There is clearly room here for a third possibility, namely, that *all physical objects* are conscious. Such is the idea behind

5.8 Panpsychism

This is the doctrine which claims that *everything* is, at least to some degree, conscious. At the present time it is not fashionable. Indeed, the OCM lists it under "animism", which is defined either to be a belief held by primitive man, or as the doctrine that *there is a spark, or germ, of consciousness present in all things.*

In the previous section we have introduced this idea in such a way that it is seen to be at one extreme of a scale, in which dualism is at the other end, with materialism at the centre. In fact, it is possible to argue that the central position cannot be sustained and that materialism must inevitably become panpsychism. The argument begins by asserting that a proton, say, is not conscious, but that a system, e.g. myself, containing about 5×10^{28} nucleons is, or at least can be. There will then be some number of nucleons, between one and 5×10^{28} at which a system changes suddenly from being not conscious to having the possibility of being conscious. This seems very implausible. The number would be a remarkable new constant of nature. Presumably it would have to be very large, otherwise we would certainly be in the situation where everything except very tiny, microscopic, systems were conscious, which would essentially be equivilent to panpsychism. Then it appears very hard to see how consciousness could exist for some number, N, of nucleons, but not for $N - 1$. We would not expect to see such a *fundamental* difference to come with such a relatively small change in the number of nucleons; nature tends not to work like this. We would much sooner expect that the changes were continuous, i.e. that there were degrees of consciousness, and that the degree allowed increased steadily with the number of nucleons. Essentially all systems would then have some possible degree of consciousness, which is the basic idea of panpsychism.

We can put the above paragraph in a slightly different form by noting that, if we accept materialism, then some physical states of my body (i.e. of *me*) are "happier" than others. The same thing would be true if you removed a few molecules from my body. (Of course if you removed too many, then more of the states would be "unhappy"!) There does not seem to be any reason to believe that, "suddenly", this process would cease and that all states would become equally happy, which would imply that consciousness had dropped to zero. Materialism has the great benefit of operating within physics, i.e. within the world that we seem to know something about. This of course carries penalties, and one of these seems to be that the general continuity of physics does not fit in well with the idea that consciousness suddenly switches on. In the words of David Griffin (1986): *The difference between the proton and the psyche is one of degree, not of kind (in an ontological sense). One who holds otherwise is a dualist, however odious such a description may be.* Apart from a technical quibble dealt with in the next paragraph, it is hard to dispute this statement.

The technical point is an interesting one concerning nomenclature. When I speak of "me", for example, I am referring to millions of possible states of the system which is me. Thus, me with my hair as it is, and me with my hair combed, are different, but are still called "me". Equally me when happy and me when sad are different physical states which have the same name. On the other hand, a "proton" is a name used for a *unique* physical state. If we change the state then in elementary particle physics we give the system a different name. The fact that these systems are so "simple" means that there are not many states available, in contrast to the situation with a man, so this is possible. Hence a "happy proton" might be, for example, the well known (to particle physicists) Δ-resonance.

It is also possible to consider panpsychism from a very different, dualist, position. Here we would assert that no physical system is conscious, and that consciousness is something outside physics. We know that it somehow becomes associated with *particular* physical systems, e.g. people, but again we have the problem of why some and not others. It is probably easier here than in materialism simply to *assert* whatever answer we choose, and that could be the answer given by Descartes, only people, or it could be people and animals, or, if we wished, the answer of panpsychism, everything. The reason why it seems easier to make an arbitrary choice here than in materialism is of course a reflection of the fact that everything is mysterious, so one more mystery does not trouble us!

Another approach to this issue is possible within dualism. Consciousness does not really "belong" to any particular physical system. It is a non-physical "thing" which under various circumstances can interact with physical systems. So the degree to which a system is conscious is the degree to which it is sensitive to the particular form of this interaction. Again such ideas suggest degrees of consciousness, and would be consistent with panpsychism.

There is a further philosophical scheme in which panpsychism is introduced in a natural way. We have already mentioned this in section 2.2. It is due to A N Whitehead and is called

5.9   Process philosophy

There are several reasons why I wish to mention some of the ideas of Whitehead. He was a mathematician and, particularly in collaboration with Russell, had a considerable influence in the development of the ideas leading up to Gödel's theorem which we discuss in chapter 9. In his later life he wrote extensively on metaphysics and the philosophy of science, being particularly concerned with the gulf he saw between the "mental" and "physical" worlds: *For some, Nature is the real mere appearance and mind is the sole reality. For others, physical Nature is the sole reality and mind is an epiphenomenon* (Whitehead, 1934). His reaction to this was to claim that *neither physical Nature nor life can be understood unless we fuse them together as essential factors in the composition of "really real" things ....* He thus rejected materialism, idealism and also dualism.

In their place he first emphasised that "change", or "process", rather than things that belonged to particular instants of time, are the key ingredients of the world. But apparently physical processes have no aim or purpose; no meaning can be attached to them: *sense perception discloses no aim in Nature.* On the other hand human experience suggests that aim and purpose are essential to understanding the world. (We explain the artifacts of man by saying *what they are for.*) Thus he met the standard dilemma, which he expressed in the words: *How do we add content to the notion of bare activity?*

His answer is to claim that, in fact, aim and meaning are present, at least in a primitive form in all processes:

> *In so far as conceptual mentality does not intervene, the grand patterns pervading the environment are passed on with the inherited modes of adjustment. Here we find the patterns of activity studied by the physicists and chemists. Mentality is merely latent in all these occasions as thus studied. In the case of inorganic Nature any sporadic flashes are inoperative so far as our powers of discernment are concerned. The lowest stages of effective mentality, controlled by the inheritence of physical pattern, involve the faint direction of emphasis by unconscious ideal aim. The various examples of the higher forms of life exhibit the variety of grades of the effectiveness of mentality* (Whitehead, 1934, p.94).

I do not fully understand what all this means. Presumably Whitehead is claiming that simple systems always, or almost always, obey the laws of physics, i.e. those laws which are not "mental" laws. The latter, however, become relatively more effective as we move to the higher forms of life. What is not clear (to me) is how the laws are to be distinguished, or why, once we have distinguished them, we are not in the position of a dualist. (Except of course that we are not following Descartes in denying the dual nature to all but humans.)

What is especially interesting here, from our point of view, is that there are some aspects of these ideas which seem to be echoed in the tentative ideas we shall suggest later, based on an interpretation of quantum theory (see chapter 12).

5.10   Summary

The central, recurrent, theme of this chapter has been whether conscious mind can be understood through physics. Our conclusion is that we would be mistaken to jump to any conclusion until, and unless, we understand the terms better. In particular we would have to try to define the word "physics". This, however, is probably a foolish thing to attempt. The history of physics suggests that physics continuously incorporates *new* things. To try to define it would be to restrict its scope. I have the impression that those who confidently deny physicalism, and those who, with equal confidence, adhere to it, have a different idea of what the word means. Alternatively they are stating positions based on prejudice, not reasons.

It is clear that physics, as we use it at the present time, does not contain the things of conscious mind. Thus it is necessary that we add these things to the world of physics. In this sense we are using dualism (like reductionism in section 2.3) as a *methodology*, not necessarily implying that consciousness will always be separate from the physical world. We *naturally* consider our minds to be something outside of the physical world and yet as interacting with it. This could be a useful working hypothesis to be abandoned only in so far as we can show it to be false, either because of some experimental evidence, or because we can show that the type of interaction between the mental and physical that is required is impossible.

At the same time we should endeavour to enlarge our understanding of what constitutes physics, and of what it can explain. We could try to answer Pippard's (1988) claim that consciousness can never be part of physics because it is not in the public domain, by saying that when it *has become* a part of physics then it will be in the public domain. Of course we may instead reach an impasse, and come to understand better the reason for physics being forever limited.

Finally, we must refer forward to our discussion of quantum theory. This affects the issue in several ways. It makes it much harder to know what we mean by "physics". *We do not understand even "simple" things like electrons,* so there is not a world of understood physics! Most efforts to provide an understanding seem to open the world of physics to something outside of itself. There is also the striking fact that quantum theory has caused us to abandon forever the idea that "real existing things" are associated with a particular point or region of space. We do not necessarily have to worry about such things as where consciousness is.

This chapter has been about opinions and contrasting prejudices; there have been very few facts. This is why the subject is not really a branch of science; there we try to solve controversies by means of experiments. It is appropriate, therefore, that we should now consider some of the experiments that might have something to say about the issues of this chapter.

Chapter 6

Experiments relevant to conscious mind

6.1 Science needs experiments

Although we have described the remarkable progress in our understanding of the physical world mainly in theoretical terms, it has always been led and inspired by experimental results. The standard model was built on the hard work and skill of many experimental physicists. It may be true that, if we had been clever enough, we could have worked it all out ourselves; we may live in the *best* (Voltaire, 1760), or the *simplest* (Squires, 1981) or, since "uniqueness" has been an elusive goal of many models, even the *only* possible world, so that we did not really need experiments to discover its properties. In practice, however, this is not how things have happened; physics has always been driven by experiments.

It is natural, therefore, that we should look to experiments to guide us in our attempt to understand conscious mind, and indeed that we should blame our lack of understanding on the paucity of results which have anything useful to say on the topic. In this chapter we shall review, very briefly, some of the experimental information that is available in relevant areas. We begin by looking at experiments on the psychology of animals. Then we describe the remarkable phenomenon of seeing without awareness. Next we describe a few elementary properties of the human brain, which lead us to consider experiments on the relation between conscious experience and various things in the brain. Finally we enter the controversial area of so-called "paranormal" effects of consciousness.

6.2   Consciousness in animals

In section 3.2 we already saw that it is unlikely to be possible to devise experiments that will tell us whether or not any creature is conscious, at least not until we are more clear about the meaning of the word. Of course this fact does not discourage psychologists from doing interesting experiments which, at least, look for signs of consciousness.

As an example there are experiments (Benninger *et al.*, 1974) designed to see if rats have any awareness of what they have just done. They are given four "options" (selected from things they naturally tend to do), namely, washing, sitting up, walking, and doing nothing. Following the action they must press one of four levers, each associated with one of the actions. If they press the correct lever then they are rewarded with food. Apparently they are able to learn to do this successfully. Clearly then the rats have memory of what they have just done, clearly also they are able to learn that particular sequences of action produce food. What, however, does this tell us about the rats' awareness? Weiskrantz (1987) gives further discussion of this experiment.

We can also do experiments that attempt to test for "self-awareness", which is an important aspect of consciousness. A baby, even as young as 18 months, is able to recognise that his image in a mirror is indeed an image of himself. The same thing is apparently true of a chimpanzee, which will relate the image of a spot on its face to the actual spot. For most animals, however, there is no sign that they have any recognition that the image is an image of their own self; they appear to be unaware of such a concept. (See Humphrey, 1986, and Premack and Woodruff, 1978.)   Are we here seeing evidence for degrees of consciousness, or do these experiments merely give us another way of measuring intelligence?

Of course the strongest evidence for consciousness in animals is that they behave in many ways very like humans. It must be realised, however, that if we adopt any sort of epiphenominalist understanding of consciousness, in which we say that it has no, or at least very little, effect upon behaviour, then we greatly weaken this argument. If I see an angry dog approaching (and dogs, uniquely in the animal kingdom, seem to be often angry), I take evasive action. So will a sheep. Now I explain my action by saying that I do not wish to be hurt, so it is natural to suppose that the sheep behaves as I do for essentially the same reason. If however we follow Descartes and assert that sheep are not conscious, then we must not do this. The light waves from the dog pass through the eyes of the sheep, onto its retina, where they are transformed into electrical impulses that trigger particular neurons that cause the sheep to run away. There is no conscious awareness of danger. At first sight this seems implausible, but of course if we are to accept any sort of *physical* description of consciousness, even what happens in the case of my own reaction, can be reduced to this sort of description. Is *awareness* of danger necessary for my action, or would the action happen regardless of my being aware? The next section describes a rather remarkable result which might have some relevance to this question.

6.3   Blindsight

Weiskrantz and collaborators (Weiskrantz *et al.*, 1974, Weiskrantz, 1987) have performed a series of experiments with brain-damaged patients, who have had lesions in that part of the brain associated with visual information from a particular region of the visual field. As might be expected the patients claim that they are blind to objects placed in that region. In particular they are not aware of a flashing light. However, when asked to make guesses about whether or not a light flash occurs in the blind region they are able to give a correct answer. They are not aware that they are able to see anything, and so believe that they are playing a guessing game. They are amazed when told that the evidence of their correct guesses shows that they can still "see", in the sense that their brains are receiving and processing visual information.

This seems to be an example of where we make the response to a signal exactly as if we were aware of it, even though we are not, in fact, aware of it. Unfortunately nobody seems to have done an experiment to see if these people would run away from an angry dog, which is only visible in their "blind" visual field. If they did, it would be interesting to know how they would then explain their actions.

Later, we shall see how other experiments show evidence for similar sorts of non-conscious processing of visual information. First, we look at the human brain.

6.4 The brain

As we noted in the opening section, this book is not intended to be about the physical mechanism of the brain. Nevertheless, to some extent this mechanism must be relevant, so it is useful to consider a few elementary facts. Actually, although a lot of detailed information is available, we do not even know the answers to many simple questions like, for example, how the brain stores information. This is not really surprising in view of one fact that is known, namely, the awesome complexity of all brains. The basic cells of the brain and of the central nervous system are of a special type called *neurons*, and there are around 10^{11} neurons in the human brain. These neurons are varied in form, but typically consist of a central *soma* attached to a long fibre called an *axon* and to several different fibres called *dendrites*. The fibres are connected to other neurons by tiny junctions called *synapses*. There are of the order of 10^{14} such synapses in the human brain, with as many as 10^5 being on a single neuron.

The neurons respond to and transmit electric impulses, the latter process being referred to as "firing". At some of the synapses the effect of an electric impulse is to release a fluid, which then alters the rate of firing of subsequent neurons. Information storage and processing in the brain seems to be concerned with the rate of firing and with change in this rate. (This is unlike the binary, off–on, situation in a computer.) All the sensory information the brain receives, and all its thinking processes, are related to variable rates of firing of neurons. We have little idea of how all this happens, but it is known that particular regions of the brain are associated with particular things. For example, we know the regions that are relevant to visual and auditory sensations. Similarly, it is possible to associate "prickly" and "burning" pains with different parts of the brain.

There are reasons to believe that conscious sensation is a result of things happening within the brain, and is not directly linked with the external cause. For example, we are conscious of our finger being pricked *only* because certain signals are transmitted from the finger to the brain; the finger itself does not communicate directly with consciousness. The evidence for this comes from so-called "phantom limb pains", where it is possible to feel pains in limbs that have been removed by amputation. Actually such results only show that the brain is capable of having the experience without the limb, it does not *prove* that the limb cannot have the experience without the brain. The fact that the pain can be felt when the appropriate region of the brain is artificially stimulated also supports the idea that the source of the sensation lies in the brain.

The next thing we note about the human brain is the remarkable fact that it is split into two roughly equal hemispheres. This has led to some exciting experiments which deserve a new section.

6.5   Split-brain experiments

The basic functions of the two sides of the brain, for most right-handed people, are shown in figure 6.1. The right hemisphere controls the left side of the body and receives visual signals from the left of the line of vision of the eyes. The left hemisphere is similarly associated with the right-hand side. In most right-handed people the left hemisphere also controls the speech, and for this reason is called the dominant hemisphere. We are normally totally unaware of this division of responsibility between different parts of the brain, a fact which is perhaps not surprising in view of the large number of nerve fibres in the *corpus callosum* which connects the two hemispheres. For example, when I look at the centre of a line of print there is no apparent difference between the way I perceive the two sides. Similarly, moving my right leg is not different in any fundamental way to moving my left leg. To the conscious mind the brain is a unity.

We might expect the situation to be very different in people who have undergone "split-brain" operations. In these operations, which are performed in order to cure certain types of epilepsy by preventing the spread of abnormal neural excitations, the connections in the *corpus callosum* are severed. What does this do to the unity of the conscious mind? The first thing to be noted is that, after a period of settling down following this operation, neither the patient nor an observer would, in the normal course of living, be aware that anything had happened. Controlled experiments, however, reveal some startling effects. If objects are shown to the right of the visual field, so that their images are transmitted to the left hemisphere, then the person can correctly see them and say what

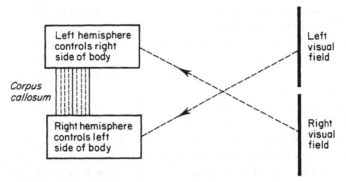

Figure 6.1. Schematic diagram showing the functions of the two brain hemispheres. The left hemisphere is dominant, controlling speech, in most right-handed people.

they are. On the other hand, if they are placed to the left of the visual field then the person will deny having seen them. Amazingly, however, he can use his left hand to write a description of the objects, which he still denies that he has seen! It is almost as if we have two people, or perhaps we should say two consciousnesses, in one brain.

It is clear that any finding of this nature would be very important to our understanding of the conscious mind. If, by the physical operation of severing nerves, we could make two conscious minds out of one, then we would have learned something of significance. Of course this interpretation is not a necessary consequence of the above results. We know that the brain carries out many operations unconsciously, so the idea of a speechless consciousness associated with the right hemisphere is not forced upon us. Indeed, it seems that attempts to demonstrate in a more compelling way that there are two conscious minds at work have proved negative, or inconclusive. The reason, or part of the reason, for this *could* be that, even with the *corpus callosum* cut, there are still many other physical connections between the right and left hemispheres. In any case the remark of one of the people subjected to these tests: *Are you guys trying to make two persons out of me?* suggests that consciousness does not "split" readily! There is much more to this fascinating topic and readers should consult MacKay (1987), from where I obtained the above quotation, for further details and discussion.

Although there is surely room for more experiments, and debate, in this general area, it is unlikely that it will be any easier to find acceptable criteria for a person to have two consciousnesses, than it is to define criteria that tell us whether an object is or is not conscious. The discussion of section 3.2 showed that this is difficult, or perhaps even impossible.

In a recent article, Swinburne (1987) uses the fact that the physical operation of partially splitting the brain gives rise to a confusion in what has happened to the conscious mind as an argument for some type of dualism. He similarly uses the even greater confusion that would arise if we considered the (at present speculative) possibility of transplanting the two halves of the brain of one person into two other bodies. Who would be the original person? We can envisage four possible answers: both, neither, the one with the left brain or the one with the right brain. He argues that, regardless of how much we may know about the details of the transplant operations, it is quite impossible for us to answer this question. In other words, he would say, the question has no meaning in purely physical terms. Since, so the argument goes, it must have an answer, these terms are inadequate. The confusion can only be resolved if the conscious mind is totally independent of the physical brain. Presumably it must choose to go to one or the other. (I must admit that I find even this resolution puzzling. It just is not possible for me to imagine myself making a choice of to which physical body I attach myself.)

6.6   The effects on the brain of mental events

In chapter 5 we met several competing theories about the nature of conscious mind. When this happens the normal scientific way of deciding between them is to do experiments. Not surprisingly, it is in the attempt to propose, to perform and to analyse suitable experiments that we realise that questions we want to answer are not properly posed. A central issue of this book is whether the world of physics is causally closed (see section 7.4, etc), or whether there are non-physical, mental, events which can alter it. In particular we want to know whether such mental events can affect the physical state of the brain.

This question is represented schematically in figure 6.2, which is a simplified version of a diagram originally produced in 1984 at a workshop in Alpbach, a holiday resort in the Austrian Alps. In this diagram the mental–neural events (MNE) are those brain processes which are in some way closely related to conscious thoughts. Indeed they are the events which according to the identity theory *are* conscious thoughts. The neural events (NE) are the other brain processes. The key issue then is whether the set of NE and MNE is a closed physical system (we are here ignoring external physical influences), or whether it can be influenced by purely mental events (ME). Obviously the two pictures, (a) and (b), correspond to the two rival answers: the identity theory, for example, in which the answer is no, and the dualist-interactionist theory in which it is yes.

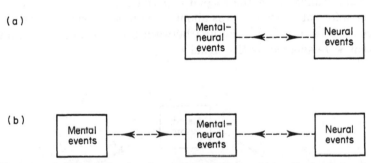

Figure 6.2. Diagrams showing opposing views of how the brain works. In (a) the brain is divided into two parts: normal neural events and mental–neural events, with the latter being identical to mental processes. These interact but, apart from physical sensory input, form a closed system. In (b) there is "something else", namely, mental events, which can interact with some of the neural events.

At first sight it now appears to be very easy to test between these two hypotheses. Can I, by a purely mental process, i.e. by thinking, cause changes in NE or in MNE? Eccles (1986, 1987) quotes two experiments in which it is claimed that this effect is indeed present. In one of these (Roland and Friberg, 1985) it was observed that there is increased blood supply, and hence increased neuronal activity, in a certain region of the brain when a person thinks about doing a previously learned complex task, without any actual movement. In the other, the subject of the experiment attended to the operation of giving a just-detectable touch to the finger, and it was found that, *even before the contact occurred,* there was an increase in neuronal activity in the area of the brain that normally receives signals from the finger.

Interesting though these experiments undoubtedly are, it is just not possible to regard them as giving conclusive evidence that mental events are causing physical events. In the first experiment, the subject had at some stage been instructed to think about the task. This instruction was given through physical channels and the whole process could therefore have happened in accordance with figure 6.3(a), rather than 6.3(b). Even if there is a lapse of time between the instructions being given and the thoughts actually starting, we do not know if the sudden decision to start to think about the task originates in a mental process or in some physical process in the brain. Similarly, in the second experiment, the subject received the information that he was about to be touched visually, so there could be a direct physical link between the eyes and the neurons that were affected by the signal.

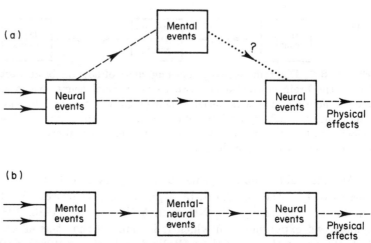

Figure 6.3. In (b) we see the claimed influence of mental events: an intention produces a physical effect. However, as seen in (a), there could be a direct physical cause of the output.

A controversial set of experiments on the timing of sensation is due to Libet *et al.* (1979, 1983). In these experiments a small stimulus (touch or electric shock) is applied to the hand of a subject who records the time when he becomes aware of the stimulus. Typically, the time recorded is about one tenth of a second after the stimulus. However, the authors claim that it takes more like one half of a second for the physical stimulus in the brain to reach what they call "neural adequacy" for the stimulus to be experienced.

Naturally these results have aroused controversy. Honderich (1987) gives references. It is not fully clear to what the results actually refer because there is a suggestion that the timing of the sensation is related to some sort of "apparent time" rather than an actual time. Also, the notion of neural adequacy seems to be very vague. Even if we accept the results, it is very uncertain what they imply. Are we intended to believe that the conscious mind can experience the sensation directly from the physical signal at the hand, without having to go through the brain? Alternatively, are the experiments intended to supply evidence that conscious mind can play tricks with time? If this is simply that we can be *mistaken*, i.e. think time passes slower or quicker in different circumstances, then the result does not seem to be very significant (for what appears to be a different view, see p.362 of Popper and Eccles, 1977). If, on the other hand, it is being claimed that the conscious mind can "foresee" the physical event then the result would indeed be exciting. However, it is hard to see how the reported results justify such a conclusion. In general, it seems as though the extremely rapid transmission of physical signals in the brain is likely to make all experiments of this type difficult both to perform, and to interpret.

In section 10.7 we shall see that a certain interpretation of quantum theory suggests a novel way of looking for evidence of non-physical conscious mind. A discussion, and criticism, of earlier attempts to give experimental support to the dualist theory can be found on p.471 of Gregory (1981). What all these experiments *do* illustrate very clearly is that it is not easy to obtain experimental evidence relative to our basic questions. One necessary step is to formulate the questions properly. The difficulty of answering them by experimental means may, in part, be due to the fact that we do not know how to do this.

6.7   The effects of consciousness on the external world

The discussion of the last section shows that experiments to determine whether purely mental events exist, by seeing if they can have an effect on the physical brain, are completely inconclusive. The brain is such a complicated system, with so much going on in it, that it seems unlikely that this situation will change. It is just not possible to take a brain as a physical system, predict what will happen, and then compare this prediction with what happens when the person starts to think.

A much simpler question, both to pose and to answer, is whether mental states can influence some elementary, external, physical system. Roughly speaking, can I move something by thinking about it? More precisely we can consider a physical system which is isolated from all external physical influences. This will behave in a predictable, calculable, manner (because of quantum effects, see chapter 10, we would have to average over many similar systems, but this does not affect the argument). The question then is whether it is possible to alter this behaviour by "mental activity". Of course, even if we found a positive answer to this question, it would not necessarily imply a dualistic model for the mind. It could be that the effect was due to some non-local influences of the neural events associated with the mental process. However, it would seem natural to limit the effects of the physical brain events to having only (known) physical influences, so there would certainly be good reasons to suspect "mental events" as the cause of any extra-physical influences.

Conversely, we might even *expect* that a "non-physical" conscious mind should be able to affect objects external to the brain. Why should something non-physical be limited by the existence or otherwise of physical connections? If my conscious mind can have an effect on the atoms in my brain, why cannot it also have an effect on other atoms? Since we do not have any idea of the nature of the supposed link between the conscious mind and the brain (see section 5.6), we cannot answer such questions. Of course there has to be some way in which my consciousness is *more* associated with my brain than with anything else; it tells me that *I* consist of this body.

In raising these questions we enter the area of the so-called *paranormal*. This is a vast, and highly controversial, area. It ranges from serious research, through cheating and fraud, to the completely foolish (perhaps mainly harmless folly but I am not completely convinced of this). There are several types of possible phenomena which it covers. *Telepathy* is the communication of information from one mind to another without any physical medium for the communication. *Precognition* is prediction of the future by "seeing" the future rather than making estimates based on present knowledge. *Clairvoyance* is the ability to "see" what is happening in distant places which cannot be seen by physical means. Finally, *psychokinesis* is the effect we were discussing in the last paragraph, the ability to affect physical objects without using any (known) physical force.

We are here ignoring even more bizarre, "psychic", phenomena like astrology, palm-reading, etc, which are one further stage removed from being plausible. It is not entirely unreasonable to suppose that the future "exists", in which case we must allow the possibility that there are ways of "seeing" it, but it is quite another thing to say that it is connected with the positions of the planets or with the creases on my hand!

A major problem we meet in discussing the paranormal is that a large percentage of the population are apparently convinced of the existence of various phemomena for which science has no explanation, even though, in the majority of cases, they have no rational ground for their belief. Even eminent psychologists are convinced. For example, Eysenck (1957) writes:

> *Unless there is a gigantic conspiracy involving some thirty university departments all over the world and several hundred highly respected scientists in various fields, many of them originally hostile to the claims of psychical researchers, the only conclusion the unbiased observer can come to must be that there does exist a small number of people who obtain knowledge existing either in other people's minds, or in the outer world, by means as yet unknown to science.*

Part of the reason for this widespread belief is perhaps the anti-science culture to which we referred in section 2.2. Part is the fact that we are unduly impressed by "coincidences". When strange things appear to work we notice them, and record them in our mind. When they do not work we fail to notice anything. (Conversely, with real things that *do* work the opposite is true, e.g. we only think about the electricity supply when it is cut-off.) Other reasons there must be, and it is an interesting study in sociology to ask what these are. The thing that concerns us here is that, whatever other reasons there may be, *evidence* is not one of them. This we must make absolutely clear: *there is abundant evidence that, at least to some sort of "first approximation", paranormal effects do not exist.*

We can do an experiment. On my desk, I place my hand. Next to it, I place a pencil. The pencil is much lighter than my hand. Nevertheless, although I can easily move my hand by a simple act of will, I cannot move the pencil (except of course by using my hand, i.e. by using a physical link). I have just tried it. Why not try it yourself? (Most readers will not bother because we just know we cannot do it.) Indeed the fact that we cannot do things like this is crucial to science. It is one of the implicit assumptions (like locality) which we employ in trying to make sense of the physical world.

Now, just because I (or we) cannot do something does not mean that it is impossible in principle. It could be that we have merely failed to learn how to do it. I am inclined to think, however, that, in this case, if it is possible then we *would* have learned, because such an ability would seem to have pretty clear evolutionary advantages (in contrast, for example, to waving our ears, which I am prepared to believe we could do if we took the trouble to learn how).

There is no need to use as measuring device for psychokinesis something as crude as a pencil. Precise experiments to measure the force exerted by conscious thoughts are possible. They show a zero effect. Hansel (1980) writes:

> *During a lecture given at Manchester University in 1950, Rhine was asked whether psychokinesis could not be measured directly with a sensitive balance. He replied that it was a good suggestion and that they might get round to trying it sometime. After sixteen years of research and after the same question must have been asked countless times, such a reply was hardly satisfactory. The plain fact is that if a direct measurement is made of the psychokinetic force by any known means, it is found to be zero.*

It should be noted that Rhine, the lecturer referred to in the above quotation, had claimed evidence for psychokinesis on the basis of ability to influence the fall of dice, an effect which surely requires quite a large force. The history of physics has not always supported Rutherford's infamous remark: *if your experiment needs statistics, then you ought to have done a better experiment,* but here at least it contains much truth. If *one* person could influence *one* balance, in *one* experiment, the point is proven, and it would be the most significant experiment in the whole history of science! So far, it has not happened.

We can similarly dismiss precognition as a serious property of the mind. Many millions of people participate in gambling, whereby they try to guess the outcome of various sporting events. If some could "see" the results before the events take place, they would rapidly become very wealthy. Given natural human greed they would not be able to restrain their earnings, and we would know about them. They do not exist.

It is sometimes implied that the theoretical "evidence" against paranormal effects is so overwhelming that we would need extremely good experimental evidence in its favour to convince us of its existence. In fact the opposite is true. There is no such thing as theoretical evidence. (The purpose of a theory is to explain things as they are. If our present theories cannot accommodate some new piece of evidence then we will happily change them.) It is the experimental evidence that is so overwhelming. I should, of course, qualify this remark by the fact that so-called "experimental" evidence always requires some type of theoretical input in order to give it meaning.

We must now mention a few, perhaps minor, but perhaps very significant, qualifications to all that we have just said. The first is to note that there is a vast amount of documented evidence for paranormal effects, allegedly under scientific, controlled conditions. Most of it is statistical in nature; not all subjects can do what is required, and those that can only have success rates that differ by a small degree from "chance". One general feature of the experiments, which is curious, regardless of whether one believes there are any real effects or not, is that as the number of tests increases the magnitude of the effect diminishes, although its statistical significance remains roughly the same. With a genuine effect we would naturally expect the magnitude to remain the same and its significance to increase. (The same thing would be true if cheating was involved.)

The fact that there *is* evidence, which *is* statistically significant, requires explanation. If we wish to deny its significance, i.e. to explain it away, then we seem to have two possibilities. One is deliberate fraud, and there are examples from several branches of science which demonstrate that this can happen. The other is the "file-drawer" effect. Suppose, for example, I wish to test for the ability of people to score higher than average by influencing otherwise random throws of a dice. I find that some people do indeed obtain better than average scores. I then send the others home and try again with the "good" ones. If I then get a roughly average performance I decide that this particular experiment was not a good way of doing it. I therefore change the conditions a little and start with a new set of people. Again I select the good ones, but this time I find that when I repeat the test with them they again do well. These are the results I publish, and they strongly support the existence of a real effect. When combined with the published results of several other similar experiments, done in the same sort of way, the whole set acquires a high degree of statistical significance. It could well be however that if we included all the data, i.e. that of the rejected people and the rejected experiments, then we would obtain an average result which is statistically compatible with no effect at all. This discussion is all somewhat oversimplified, and it must not be thought that workers in the field are not aware of the problem. Many of them are, and efforts have been made to overcome it. Good discussions, from rather different viewpoints, are given in Hansel (1980) and in Radin and Nelson (1988).

The second qualification concerns the fact that on any reasonable guess at the theory that might lie behind such effects, they would be expected to be elusive, and hard to repeat. For example, in no branch of science do we ever exactly repeat an experiment; we can only try to make the relevant conditions the same. This requires that we have at least a general idea of what is going on; otherwise we do not know what is relevant. In physics this condition is usually satisfied; in psychic experiments it never is! A related problem is the non-locality, perhaps in time as well as space, of consciousness. There are several aspects to this. There is the simple fact that something non-physical does not seem to have a position. Then, and particularly if we are somehow to relate consciousness to quantum effects as we shall try in chapter 11, there is the curious non-locality of quantum theory as discussed in chapter 10, etc.

What this might mean is not clear, but it does seem as though we might well have difficulty in knowing *which* dice, or whatever, we are trying to influence, or when. Further, it seems clear that if we wanted to influence a dice to score above average we would have to have information on how it was falling, so that we could predict how it would land and then see if, by a small couple in a suitable direction, we could make it land so as to give a bigger score. In other words, even if we knew how to exert a force on a dice, this would not obviously help us to achieve a better than average score. In fact the subjects do not actually see the dice thrown, so what is it that interprets their wish to get a six, into some force on the dice? Even if they had a specific physical way of influencing the dice, it would still be extremely difficult for them to utilise this force to obtain any given effect on the numbers.

It is possible to regard these difficulties as helping us to understand why evidence for the paranormal is so insubstantial. Certainly there are several examples in the history of physics where it has been seen to be wrong to interpret "no evidence for" as "does not exist" (parity violation, CP violation and neutral currents are recent examples). On the other hand, it is perhaps equally reasonable to regard them as supporting our mistrust in the reliability of the above experimental results.

We close this section by looking in a little more detail at one particular set of experiments which apparently show that it is possible to influence otherwise random quantum events. There are three reasons why we might expect this type of experiment to be the most promising place to look for evidence of psychokinesis. Firstly, the effects concern microscopic phenomena, so the objection that if they exist they would have been seen, for example in an experiment with a sensitive balance, does not apply. Secondly, and again because the effects are so small, there does not seem to be any obvious evolutionary advantage in our acquiring the skill, so it is no surprise that we are unaware that we have it, and that the skill is only latent. Thirdly, as we shall see in more detail in chapter 12, there are theoretical reasons for suggesting that the randomness inherent in quantum observations might, to some extent, be present simply because we fail to utilise the power we have to "choose". A deliberate attempt to choose the results of a set of experiments might give a different probability distribution to that predicted by quantum theory.

As an example of the sort of effect we are thinking of here we might consider the decay of a radioactive atomic nucleus. The quantum theoretical description of such a system is a wavefunction that contains a growing probability for the particle to have decayed. The theory does not predict a particular time when a given particle will decay. We shall suggest in section 12.4 that a conscious mind might be able to exert an influence at this point, e.g. by increasing the chance of the decay being observed. If this is possible then we might expect to be able to influence the rate of decay of radioactive substances.

The experiments that come closest to providing relevant tests are those performed at Princeton by Jahn and Dunne. Various devices are used to produce random physical processes, which are then transcribed to generate a count rate. Subjects are asked to see if they can influence this rate either to increase, or to decrease, or to remain the same. Small, but statistically significant, effects are seen, as is illustrated in figure 6.4. The results of many years work are summarised by Jahn and Dunne (1986, 1987) in the following words:

1) The calibration data conform to theoretical expectation and display no artifactual aberrations of any statistical significance.

2) The primary effect of the PK efforts is to shift the means of the binary output distributions slightly, usually in the intended directions, with little detectable changes in the standard deviations or higher moments. The statistical merit of these small shifts of the means depends on the number of trials processed, and can compound to highly significant anomalous effects over very large numbers of events.

3) Individual operators display characteristic "signatures" of achievement on such experiments. These may be dependent on the particular modes of data acquisition, but are relatively independent of the specific device or system employed.

These experiments use a variety of means for generating the random numbers, and they are not all directly related to quantum processes. Hence it is by no means obvious that, even if they are regarded as significant, they provide evidence for the sort of quantum influence we would like to test. It is puzzling why there is no dependence on the mechanisms involved. Such a dependence is

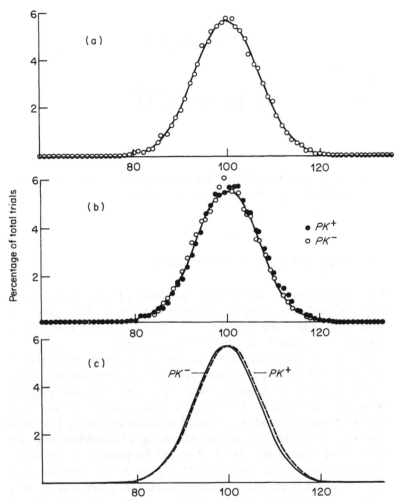

Figure 6.4. Conscious influence on random processes. (a) shows how unbiased results fit the expected curve about a mean count of 100. (b) shows the result of efforts to increase (PK^+) or decrease (PK^-) the rate. Gaussian fits to the data are given in (c).

expected because any real effect must surely operate at the stage of the physical process, and there is no reason why different processes should show effects of the same magnitude when transcribed into numerical output.

Chapter 7

Free-will

Free will cannot be debated but only experienced, like a colour or the taste of potatoes   (Golding, 1959)

7.1   A property of conscious mind

Some years ago, in a book about the story of fundamental physics (Squires, 1985), I wrote: "Freedom is a property of the conscious mind: if I *think* I am free then I *am* free." My son wrote in the margin: "if I think I am a fish, then I am a fish?"

It is a reasonable objection that he was making. Nevertheless, I believe that he was mistaken, and that the mistake is a source of much confusion about free-will. It is clear that the property of being a fish can be defined quite independently of any reference to conscious mind, e.g. through appearance, habitat or, more precisely, by an appropriate set of zoological classifications. Regardless of what I may *think*, it would be possible to verify the truth or falsity of the statement that I am a fish. However, and this is the point that it is vital to understand, free-will is exclusively a property of conscious mind, i.e. it is like red or fear; *there is no possibility of defining it otherwise*. Of course, like, for example, red, it may sometimes be possible to relate it to certain features of the non-conscious world (we consider this in sections 7.5 and 12.3), but free-will itself is meaningless outside of conscious mind.

Such a statement clearly raises some problems, to which we shall return in section 7.3, and furthermore it is obvious from the literature on the subject that it is not generally accepted (e.g. Searle, 1984, p.94). We must therefore try to justify it. To do this we have to show that there is no alternative way of defining something that we would recognise as free-will. This is the aim of the next section.

7.2   Freedom as the possibility of alternative action

When I assert that there is no possibility of defining free-will, except by regarding it as a property of the conscious mind, what I really mean is that *I know of no other way*, which is a much weaker statement. However, the attempts that others have made to find an alternative, reveal how difficult it is likely to be. To be specific I shall here examine one such attempt. In his book *Free Will*, Thorpe (1980, p.7) writes:

A decision is free if the agent could have decided otherwise.

All definitions of free-will, outside of conscious mind, seem to have a similar form to this.

The difficulty in understanding this statement lies in the words "could have". These are conditional, so we must understand the conditions. We are implicitly supposing that we could repeat the decision-making in identical circumstances and obtain a different outcome. Of course, in practice such a repetition is not possible, since nothing is ever repeated exactly, but this fact should not deter us from thinking about the possibility. As we have several times noted, truth is not necessarily restricted by what I might be able to do. To keep things simple we will divide the universe, i.e. everything that there is, into two parts: *the agent* and *the external world*. In doing this we are not necessarily implying that the division is in physical space, rather we are thinking of it in the mathematical, set-theory, sense. Also, we are willing to include in "everything" non-physical things, like thoughts, ideas, etc. The fact that there might be no natural division does not prevent us from making what is to some degree an arbitrary split.

Having made this division, we can try to make Thorpe's definition more explicit. We could say: ... *the agent could have decided otherwise if the external conditions had been different.* It is immediately obvious however that such a definition has nothing whatever to do with free-will. The most primitive system behaves differently if it is placed in different external conditions. For example, the thermostat of section 3.2 turns the heat on or off according to the temperature of its surroundings, but we do not normally describe such action as being an exercise of free-will. We shall therefore assume that the above definition can be replaced by:

> *A decision is free if, in identical external circumstances, the agent could have decided otherwise.*

The definition is now explicit about the properties of the external world. What about the agent? Is the agent supposed to be different, or identical? Let us try the first suggestion, in which case our definition becomes:

> *A decision is free if, in identical circumstances, the agent could have decided otherwise because the agent was different.*

This of course is how we normally think of free-will. We speak of Peter having made a particular choice because he is that sort of person, or we might express surprise at Mary's choice because we thought she was different. Indeed, it is through observing the sort of choices that people make, that we determine to ourselves what sort of people they are. However, it is important to realise that what we are observing here is the obvious fact that different systems behave differently. This point becomes clear if we realise that the above "definition" would give free-will to all systems. For example, although a particular thermostat might turn *off* the heat with an external temperature of 30°C, a different thermostat, i.e. one set differently, would turn it *on* at the identical temperature. Neither would be said to be making a free choice; they would be acting differently because they *are* different. If I wish to regard free-will as something that I have, but a thermostat does not have, then I cannot use this definition.

There remains one further possibility for making Thorpe's definition reasonable, namely:

> *A decision is free if an identical agent, in identical circumstances, could have decided otherwise.*

This of course means that the complete system, agent plus external world, is not deterministic, i.e. what happens is not a consequence of the initial conditions but is chosen randomly. This may be true of reality, but again it has nothing to do with what we mean by free-will; indeed it is exactly the opposite of what we mean. When I make a free decision, I use all the available evidence (this is the effect of the external world) and somehow process it through my brain (this is the effect of the properties of the agent) so that a verdict is reached. Nowhere in such a process do I use anything that appears to be random. Of course, I could let a random process "decide" for me, i.e. I could simply toss a coin. Then, however, I would be deliberately refusing to make a free choice.

We can summarise this discussion as follows. The statements

>*I might have acted differently if the circumstances had been different*

and

>*I might have acted differently if I had been different*

are both true, but they are still true if "I" is replaced by any object, alive or dead, conscious or not. They are simply expressions of the fact that if we do different experiments we expect to get different answers, and they do not provide a reasonable definition of free-will. The statement

>*I might have acted differently, even if the circumstances had been the same and I had been the same*

is to say that the world is not deterministic, and that some things just happen for no reason. This may or may not be true (we shall discuss it later in more detail), but again it is not in any obvious way related to free-will.

The conclusion of this section is that when we try to make Thorpe's definition precise, it clearly fails in its attempt to define free-will. I believe this will be true of all similar attempts that do not recognise that free-will is a property of conscious mind. Note here that I am not saying that when I attempt to define free-will by other means I discover that we are not free and so I abandon the definition. The claim is much stronger: free-will *is* an experience! It is indeed our certainty of the reality of this experience that makes us reluctant to see how it can be reconciled with a totally physical description of the world (see section 7.4 below).

7.3   Free-will as an illusion

If the above conclusion is correct, then we have no choice but to accept the testimony of an individual who tells us that he is free. As my son realised, this means that we have the problem of why we should believe him. We can think of several reasons why we might be reluctant to do this: he might be lying, he might be mad, or he might be mistaken. There is little we can do about the first possibility, and certainly in some cases it will occur, e.g. a child might want to impress us with his generosity and so say that he gave his pocket money to a charity, whereas, in fact, he may have been made to do this by his mother. In fact of course it is more likely that a person who *denies* freedom will be lying. Politicians, in particular, are very prone to use the expression *there is no alternative*, when clearly it is false. Such things however are of doubtful relevance to our discussion; in general, particularly where there is no motivation for not being truthful, we can ignore this possibility.

The problem of the person being mad is more difficult. Certainly this would be the explanation we would use if a person said he was a fish, against all the other evidence. All that we can reasonably say is that most people are not mad, so that although in some cases the evidence might be unreliable, in general we should believe them. We are of course again using the fact that other people are like ourselves (compare section 3.2). I do not find it surprising when a person tells me he is free, because I know that I am.

This brings us to the third possibility: I may just be "imagining" that I am free; it may just be a delusion. Again, of course, this can certainly happen in some cases: just as I can imagine the experience of red, and this is clearly different to the actual experience of red, which would be due to some particular source of light: so I can envisage "imagining" being free. Indeed, have we not all done this when we have thought to ourselves: *If only I could go into that shop and buy whatever I want?* Here we are imagining freedom. But, surely, we are only able to do this, because we have experienced the reality of freedom in some other contexts. We can, at least to ourselves, readily distinguish *imagining* freedom from *experiencing* it.

The experience of being free is a real experience. It is more surely real than physics, than quarks and leptons, than the big bang, than you, indeed than all the things outside of my mind. When we have the experience of freedom, we recognise it. When we lose it, we long for it. It is sometimes a source of pleasure, sometimes of pain, but it is an important part of what we are. If I am real, then my freedom is real too.

7.4   Free-will and determinism

The relation of free-will to determinism seems to cause much confusion. Whilst it is readily accepted that behaviour that is decided at random is not the same as that which results from free choice, there seems to be reluctance to accept that determinism and free-will are not only compatible, but that the former is an essential ingredient of the latter. It is because of this reluctance that some writers seem to imply that there is a sort of "middle ground" between determinism and randomness. I do not believe such a mythical place exists, or that it is necessary.

Note that in the discussion of this section it does not matter whether in fact reality is, or is not, deterministic. The claim that I wish to make is concerned with logic, not with reality. It is that there is no incompatibility between free-will (in the only meaning of that expression that makes sense) and strict determinism. The best way of understanding this is to examine the reasons that are sometimes used to support the counter claim. Roughly speaking all such arguments go something like this. If **you** knew all about **me** at some time, say, t_0, and if you also knew about the external influences on me at the same time, then in a deterministic world you could predict what I would do at any time t later than t_0, and hence my freedom would be an "illusion". Although I might struggle to decide my course of action, such a struggle would, so the argument goes, be a waste of time because you would already have written down the outcome of my deliberations. Everything here is correct, except for the conclusion! In knowing everything about me, in order to calculate how I would behave, you would in fact be creating a replica of me, perhaps in your brain, perhaps on a computer, or in some other way, and this replica would have to be identical to me, in all relevant aspects. In other words, as far as the particular decision is concerned, the replica would be another "me". The fact that two "me's" make the same decision in the same circumstances is then not surprising, it is simply a statement of the fact that I *made* the decision and did not "toss a coin". Both I, and my replica, would have the same struggle to reach the decision, and we would come to the same answer. This is why I believe that some form of determinism is essential for the idea of free-will to make sense. Of course if **you** are foolish enough to tell **me** the answer arrived at by the replica you have made, then I would be free to be "awkward" and hence to make the opposite decision. It is probably unnecessary to point out to readers who have got this far, that such behaviour does not invalidate anything we have said; your replica behaved differently to me because it operated in different circumstances, in particular, it had not been told about a replica that had already chosen.

It should be noted in passing that we have here tacitly *assumed* the possibility of making a replica. Such an assumption might, however, be false. This of course would destroy the above anti-deterministic argument completely, although, at the same time, it might make it easier for us to accept that determinism does not destroy free-will. The point is that a sufficiently good replica of me, i.e. one that would always behave like me, might in fact have to *be* me. At the level of classical physics such a claim appears to be untrue; we could imagine making identical thermostats, for example. There would be practical problems in actually making them identical, but not problems of principle. In quantum theory, however, things are not necessarily so simple. The quantum world does not really have particles from which things can be made. Rather it has waves, and it is not obvious that the theory allows us to talk of two independent, identical, quantum objects. If we adopt a dualistic picture, in which conscious minds are not "made" of anything, but just are, then it again becomes difficult to see how we could make a replica of a person.

Of course, to some degree, in fact, we all quite frequently carry out the process of mentally constructing a replica. Whenever we try to work out for ourselves what a person will think, or how she will behave, or even to explain how she actually behaved, we are making, within our limited knowledge, an approximation to what we believe the person to be. It never occurs to us when doing this that we are denying, in any way, the person's absolute free-will.

The view that is being advocated here is sometimes called **compatibilism**. It asserts that there is no conflict (and I would like to emphasise the words *no conflict*) between free-will and strict determinism. A detailed defence of this view is given by Honderich (1988), and by Tipton (1988), who claims that most "analytical philosophers" are compatibilists. (Neither of these authors goes as far as I would like in accepting that free-will could not be anything other than an experience; maybe it requires the insight of a novelist rather than a philosopher to recognise the similarity between the taste of potatoes and free-will.) Those who criticise compatibilism do so, I believe, on the basis of the false understanding of free-will considered in section 7.2. Thus Searle (1984, p.89) says: *The problem about compatibilism, then, is that it doesn't answer the question, 'Could we have done otherwise, all other conditions remaining the same?'* The question has, however, nothing to do with compatibilism; it is a question about determinism and the deterministic answer is a clear "no". The *same* me, in the *same* circumstances, would freely choose in the *same* way. Such a statement is not a denial of free-will; rather it is an affirmation of the fact that when I decide my particular action then that, and not something else, is what happens. Any other behaviour would require that there are random features in the world.

Even the language we use supports compatibilism (though I doubt whether we can legitimately use this as an argument in its favour). We sometimes refer to "choosing" as "determining" what will happen. If instead, when I choose to lift my arm, it sometimes goes up and sometimes goes down, then I would have to abandon both determinism and free-will.

The reason for our difficulty with compatibilism lies, I believe, in our reluctance to regard *ourselves* as part of the world. We are willing to so regard our hands, our hair, our lips, etc, but we want to keep something back, something that is *me*. As long as we do this (and the question of whether what we keep back is something physical or not is irrelevant here), then the rest of the universe is not deterministic, because it can be altered by "me". We shall study this idea of separation further in the next section. It is the misunderstanding of this that causes us to regard determinism as somehow threatening what Honderich (1988, ch.7) calls our "life hopes". This is an empty threat. How can it be that the introduction of a random element between the world at time t_1, and the world at time $t > t_1$, contributes anything to my life hopes?

There are also psychological, or perhaps sociological, reasons why we like to think free-will actually means something different, and something that is not compatible with determinism. As we discuss further in section 13.4, we have a strong desire to "blame" people for what they are, and this leads us to attribute to free-will properties that it cannot have.

7.5   The origin of free-will

*What does choice mean in a system of causal, physical inter-
actions?*   (Blakemore and Greenfield, 1987, preface)

Several times in this chapter we have emphasised that free-will is a
property of the conscious mind like, for example, red. However, we
know that red is normally caused by something clearly physical,
namely, light of a particular range of frequencies. It is therefore an
obvious question to ask whether we can push the similarity as far
as having a similar physical "cause" of free-will. When I experience
free-will, is the experience in any way related to specific physical
processes and, if so, what are they?

It may be that the sensation is associated with activity in particular parts of the brain, or with certain specific patterns of neural
activity. It is then very obscure how these could be different to
other parts, or other activities, in such an appararantly fundamental way as to account for the difference between free and non-free
actions.

What is, at first sight, a more promising possibility would begin by supposing that, in a sense to be described below, "I" can
meaningfully be separated into two parts which we shall call I_o
and I_i. The suffices o and i were chosen because they might refer
to "outer" and "inner", but I am not sure what these terms mean
in this context. Notice that again the division is not necessarily
in physical space; it is meant in the set-theoretic sense. We then
want to be able to say that I_o normally acts independently of I_i,
but that occasionally I_i intervenes and controls the behaviour of
I_o. On such occasions we would experience free-will. The world
consisting of I_o plus all things external to me would not be deterministic, because it would be influenced by "my free-will", i.e. by
the effect of I_i. It is a model something like this that underlies
much of our thinking about the operation of free-will.

We should notice, however, that we can make models of the
above kind in a totally trivial way. For example, I_i could be my
big toe. The model then obviously satisfies all the conditions of the
above paragraph. It is clear that we need something more if we are
to understand the origin of the sensation of free-will in this way.
If I_i is distinguished by being "non-physical", if it is not made of
quarks and leptons, if it is the "other" of dualism (see section 5.6),

then the separation into two pieces becomes more significant. This is clearly the type of model which dualist theories of mind would suggest. As we have already stated such theories meet the problem of how the two parts interact. Nevertheless, the experience of free-will certainly *seems* like this; a fact which should not be ignored.

The issues of this section can be clarified if we compare the behaviour of a plank of wood, floating on the sea, with that of a ship. The former obviously has no free-will or choice in its behaviour; it goes where the currents take it. The ship, on the other hand, does not merely follow the currents; it is steered by the will of the pilot. If we consider the world of *everything except the pilot*, then it is non-deterministic. This is the picture we tend to have of people: we separate from them a "pilot", their free-will, and we then, quite rightly, assert that they are not governed by deterministic laws. But this is because we made an apparently arbitrary choice to consider just a part of the person. If we include everything, *including the pilot*, then the non-deterministic feature disappears. It seems to me that nothing here is changed in any fundamental way if the pilot is, or is not, "different", i.e. not made of quarks and leptons. In particular, there does not seem to be any reason why free-will is made more real if there is a non-physical pilot.

The example of a ship with a pilot does introduce one other feature that may have some relevance: the pilot may be in radio communication with somebody on the shore. This would imply a sort of "non-local" influence on the behaviour of the ship. Later we shall see that such things could possibly occur in the quantum world.

7.6   Purpose and design

Desire is the essence of man (Spinoza)

A traditional argument for the existence of God is that the universe is too amazing to have happened "by accident" (see the discussion in section 4.6); rather it shows the features of having been designed, a fact which of course implies a designer. The argument might refer, for example, to a watch, which could not happen by accident but which, we say, is the consequence of the work of a human designer. This argument has been the subject of criticism, mainly on the grounds that evolution, operating over millions of years, could produce effects which, superficially, *look like* design.

It is not our purpose here to enter into this particular argument. We are simply concerned with its starting point, e.g. with the watch itself. We readily and naturally accept that such things are indeed evidence for design, i.e. for the fact that people have purpose. We do not imply by this that the particles which make up the watch have at any stage failed to obey the laws of physics. Why then do we not worry about the fact that it seems to require an incredible "accident" for the initial conditions of the universe to have been just such that particular quarks and leptons finished up in the form of a watch? (see figure 7.1(a)). The reason is that we recognise that there has been personal intervention: at some stage I wanted a watch, this desire acted upon my brain, which in turn acted upon my body, which did particular physical things that caused somebody to design and make a watch. The actual physical state of the particles from which the watch was made was essentially irrelevant. Whatever had been their physical state (within some large limits), my desire for a watch would work on the physical world, through the laws of physics, to produce a watch. Thus we replace the (unlikely) scenario of figure 7.1(a), by that of figure 7.1(b), which seems much more reasonable. This illustrates the point we noted earlier, that the existence of conscious mind seems to affect the world: if nobody "wanted" a watch it seems clear that none would ever occur!

Have we solved the problem of the watch? Two attitudes are possible. We could say that the second figure (b) is actually not different to the first (a); we merely have to draw a large box, as in figure 7.1(c). Then it is still the initial conditions that have created the watch, in which case we have not really explained anything. Alternatively we could take a positive view and say that the second picture shows how the first actually works. The initial conditions of the universe created conscious minds, with purpose; then everything else is easy to understand.

Which attitude is correct? Somehow purpose seems to be such an integral part of the universe as I experience it, that I find it hard to believe that it just "evolved" from a situation without purpose. To continue such a discussion, however, would seem to go beyond our immediate concerns in this book, so, for the present, we leave it.

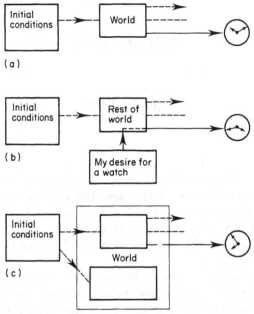

Figure 7.1. Three possible ways of understanding the existence of watches.

7.7 Summary of our conclusions about free-will

What I have been trying to say in this chapter can be summarised in the following statements. (For brevity they are stated here rather more dogmatically than they deserve.)

To experience free-will is to have free-will, because this is what free-will means. It has never meant anything else and it cannot ever, under any circumstances, mean anything else. Another person has no more right to argue that my freedom is not real than he would have to doubt me if I said that I was happy. Although there may be aspects of the world that are not deterministic, my sensation of free-will is totally compatible with, and normally assumes, strict determinism. The experience of free-will does not, in any obvious way, help us to decide between a materialistic or a dualistic view of mind. The existence of free-will is not more, or less, of a problem than the existence of consciousness.

We have not finished with all the issues raised in this chapter; many of them will occur again.

Chapter 8

Time

There are two aspects of the basic laws of physics, as we understand these at the present time, which are difficult to reconcile with our experience of the world. One concerns the measurement problem of quantum theory, a topic we meet frequently throughout this book; the other concerns the nature of *time*, and will be the subject of the present chapter.

We exist in "space–time", in the sense that we are aware of events happening at particular points of space and at particular times. Of course we have no idea what space and time "are", but maybe that is not something about which we should worry; they just *are*; they are things that cannot be expressed in any other terms. There are however several, possibly related, aspects of time which, although natural and familiar in our experience, are, from the point of view of physics, extremely mysterious.

This chapter contains a few equations but readers who ignore these will not, I hope, lose the general line of argument.

8.1 Time and the laws of physics

We begin by discussing the properties of time as they appear from the laws of physics. The first thing to note is that *physics is invariant under time reversal*, that is, the fundamental laws of physics do not distinguish a direction of time: the past and the future are treated in identical ways.

To understand what this means, and how it comes about, we first consider classical mechanics, in particular, Newton's laws of motion. The first of these states that bodies continue in a state of rest or uniform motion unless influenced by a force. Clearly such a law is independent of the direction of time, i.e. it will be true regardless of whether we take the word "continue" to mean moving forward or backward in time. The second law states that the acceleration of a particle is proportional to the force acting on it. Now the acceleration is the rate of change of the rate of change of the position and as such it does not alter when we change the direction of time (i.e. run a film backwards). This is in contrast to the velocity, which is the rate of change of position, and which clearly changes sign under time reversal. In mathematical language we have

$$acceleration = \frac{d^2\mathbf{x}}{dt^2},$$ 8.1

where $\mathbf{x}$ is the position, from which the invariance when t becomes $-t$ is obvious. Newton's third law, that action and reaction are equal and opposite, is also not affected by changing the direction of time.

We can illustrate the invariance of Newton's laws by considering the path of a projectile thrown from the surface of the earth. This path is a parabola and, as shown in figure 8.1, it is still a parabola if traversed in the opposite direction, so if we saw a film of the trajectory we would not be able to tell if the film was running forwards or backwards. Two qualifications must be made to the last statement. First, it is possible that the ends of the trajectory would immediately distinguish the two directions; if, for example, the projectile just falls to the ground and bounces, then the reversed film would appear very strange and unlikely. Also, because of air resistance, the projectile would actually lose energy as it moves along its path, so it would not exactly follow a symmetrical parabola. Neither of these effects in fact represent any violation of the reversibility principle at the fundamental level, as we shall discuss more carefully in section 8.3. Another example would be given by the collisions between the balls in a snooker game, which would obey the same rules if the velocities were reversed. Here again, of course, readers may well protest that if we ran a film backwards we would soon realise our error: the balls would speed up rather than slow down as in real life, and, even more surprisingly, the original triangle would reform!

In quantum theory, Newton's law is replaced by the Schrödinger equation, which is first order in the time derivative:

$$i\hbar\frac{\partial}{\partial t}|\psi> = H|\psi>.$$ 8.2

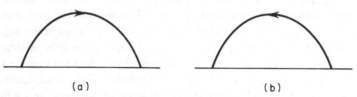

Figure 8.1. Showing how the parabolic path of a trajectory, (a), appears the same when time is reversed as in (b).

The left-hand side of this equation now does change sign if we reverse the sign of t. However, clearly, such a sign change makes no difference because it is equivalent to changing the sign of i. This in turn just corresponds to a change in the, so-called, phase of the wavefunction, and such a phase change does not alter physics, since it cancels in all physical quantities. (For readers who know about complex numbers the point is that physics always comes from ψ multiplied by its complex conjugate, $\overline{\psi}$, so a phase factor $e^{i\alpha}$ has no effect.)

This time reversal invariance of quantum theory depends upon interactions in the Hamiltonian, H, being, in the mathematical sense, real functions. In fact, however, there are some interactions which have a small imaginary part, and which give rise to very tiny effects, seen so far only in the decays of certain particles called *K-mesons*. Such small effects play no role in our "macroscopic" experience of time, so they can be ignored in the present discussion.

Of course, as we have already seen, the wavefunction is not the complete story of quantum theory, and it may be that the process of measurement (see chapters 10 and 11) introduces important effects which are not invariant under time reversal. We shall look further at this in our discussion of quantum theory, and for the present work within an essentially deterministic, classical, framework.

The second feature of time that comes out of the laws of physics is that, in many respects, it is similar to space. At first sight this seems a surprising statement. Time is "one dimensional", in the sense that we need only one number to specify the time (once we have chosen the origin and the unit in which it is measured). Space, on the other hand, is three dimensional; thus to specify the position in a room we can choose one corner as the origin and then give the distance from two perpendicular walls and from the floor.

Figure 8.2 illustrates how we might define these *coordinates* in the case of two dimensions, which of course is all that we have on a flat page. The actual value of the coordinates, for a fixed point P, depends on how we choose the axes, and two possible choices, related by a rotation, are shown in the figure. The length of the line OP is of course independent of the choice of axes. It is given by the well known theorem of Pythagoras:

$$(OP)^2 = x^2 + y^2 \qquad\qquad 8.3a$$
$$= x'^2 + y'^2. \qquad\qquad 8.3b$$

These equations express the invariance of the quantity $x^2 + y^2$ under rotations. More generally, in three space dimensions, the corresponding quantity $x^2 + y^2 + z^2$ is invariant.

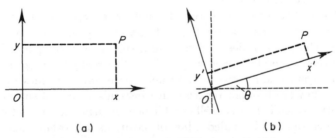

(a)   (b)

Figure 8.2. (a) Showing how a point P is determined by two numbers, x and y, when the origin and axes are given. (b) Here we see how the same point P is now given by two different numbers, x' and y', when new axes, rotated through an angle θ, are used. A little geometry shows that $x' = x\cos\theta + y\sin\theta$ and $y' = -x\sin\theta + y\cos\theta$.

Now, just as we can represent a position in three-space by giving three numbers, so we can represent an *event*, something happening at a particular point of space and time, by giving four numbers (x, y, z, t), i.e. a point in four-dimensional space–time. Clearly this is even more difficult to represent on a two-dimensional page, but we can draw, for example, an x, t plot which might show, as in figure 8.3, the way in which the x-coordinate of a particle varies with t. If we ignored the other space variables then an x, t plot shows the whole of space–time; it is the "stage", on which the universe, its past, present and future, exists. A given line on such

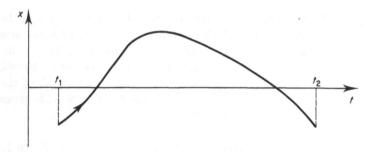

Figure 8.3. A graph showing how the position of a particle moving on a line, denoted by its distance x from an origin, varies with time, t. At first the particle is moving in the positive direction (x increases with t), but later it moves in the opposite direction. At time t_2 it has returned to its starting point.

a plot, perpendicular to the t-axis, corresponds to a fixed time, e.g. "now" (see next section).

The fact that we can represent events on an x, t plot, in a similar way to the representation of points on an x, y plot, for example, shows that we can describe time in a similar way to space. However this is merely a way of speaking. The really intimate nature of the connection between time and space only becomes evident when they are brought together in Einstein's theory of special relativity (which is now so well established experimentally that it has as much right to be called a law of nature as any other such law). As we saw already in section 4.1, this theory is based on the fact that the velocity of light is always the same, regardless of the motion of the observer. Now different observer velocities are obtained by rotating in the combined, four-dimensional, space–time coordinates (x, y, z, t), and the statement of special relativity is that the quantity

$$s^2 = x^2 + y^2 + z^2 - c^2 t^2 \qquad 8.4$$

is invariant under such rotations (cf. equation 8.3). Here time is seen to play a very similar role to any of the three space coordinates. The extra factor c^2 reflects only the fact that we usually measure time in different units to space, e.g. seconds rather than metres. The only significant difference between equation 8.4 and expressions like that in equation 8.3 is that, whereas the space coordinates appear with a positive sign, the time has a negative sign. This negative sign is the only known distinction between space and time at the most fundamental level of physics. It is a big question, to which we do not know the answer, how, and if, this sign can explain the very real differences in our perception of space and time. We shall discuss these differences in the next section. It is worth adding here that many treatments of cosmology tend to work in, so-called, "Euclidean" time, in which the minus sign is replaced by a plus sign. Surely something very important is missing in such discussions.

8.2 The experience of time

There are three ways in which the human experience of time differs from what might be expected on the basis of the above section. First, it is clear that *human experience distinguishes the two directions of time, the past and the future, in a clear and unequivocal way*; we do not confuse them. A simple expression of this fact is that we remember the past, but we do not "remember" the future. The analogue of remembering the past is predicting the future. There are of course people who claim to have paranormal powers in predicting the future. If these claims are valid (which I doubt), it would be interesting to know whether such people do this by a process which to them is akin to "remembering" the future. Another is that we believe we can influence the future, whereas we would make no claims that we could influence the past. All this of course is very familiar and obvious; we would indeed explain it as a trivial consequence of the fact that the past "has happened", whatever such a remark might mean (see section 8.4). These elementary properties of our experience become very significant, however, when confronted with the fact, noted above, that the basic laws of physics make no such distinction between past and future.

We can see the problem in another way if we compare with our experience of space. Here we agree with the expectations from physics: the positive x-direction and the negative x-direction, e.g. in front of me and behind me, are not different in any *fundamental* way. I can for example move in either direction, whereas I have no such choice of options with regard to time.

This then is our first problem: why do the past and the future seem to us to be so fundamentally different when all the basic laws of physics do not allow us to distinguish them? Is this again a remarkable feature of conscious mind, showing that it somehow is not amenable to the laws of physics? In spite of what we have said above it would be a mistake to give an unqualified affirmative answer to this question. The reason is that it is not only in human experience that there is a clear asymmetry in the direction of time, as we shall see in the next section.

The next curious thing about our experience of time is that we have a sense of "moving" through it. We cannot stop, or, as we noted above, go backwards. Again this seems an obvious property of our experience of time. This is what time *is*; it orders our experience, and it therefore makes no sense to think of going backwards; we would somehow have to "unexperience" what we had once experienced.

Related to the experience of moving through time is the concept of "now" or the "present time": the unique moment that divides the past from the future. Neither the concept of moving through time, or that of "now", have any place in physics. Indeed, as we saw in the last section, time and space are treated in a very similar way in physics at its deepest level. However, our experience of space is very different to that of time. We do not inevitably move through space, we can stand still, or go in the other direction. Similarly the notion of "here", is very different to that of "now". We somehow expect everyone to experience the same "now", but do not have the same attitude to "here". We can stay "here", but we cannot stay at "now". Again we can ask whether these things are properties of conscious mind, or whether they reflect something else in physics. Before we try to answer this we must learn about those aspects of the external world which also show a preferential direction of time.

8.3   Time asymmetry and thermodynamics

We noted above that the motion of snooker balls is not in fact quite symmetric with respect to time, because it is evident that the balls slow down and indeed eventually stop. If we took a film of the motion and played it backwards, we would see the balls spontaneously start to move! This does not happen in the real world, so we would know that the film had been reversed. Now the reason for this slowing down is that the balls, the table on which they roll and the air through which they move, have a structure; in particular they are made of atoms. As the balls collide, and roll on the table, these atoms are caused to move, thereby causing energy to be transferred from the ordered, collective, rolling motion into random motions of the atoms, motion which we perceive as heat. At all stages the time-reversal-invariant laws of physics are being obeyed; as a ball stops rolling its momentum is transferred to the atoms of the cloth and the air. If we saw the *details* of the time-reversed film then even the spontaneous motion would seem reasonable: we would see how a lot of atoms travelling in the same direction would suddenly strike the ball and cause it to start to roll. It would be quite *possible* for this to happen, in the sense that it would be compatible with the laws of physics, but it would of course be very unlikely, so unlikely that we can confidently assert that it does not happen (at least not in a world running in the normal direction).

The reason for the apparent direction in time of the processes in the world is thus that we only observe the ordered effects of the motion of billions of atoms, and the effect of interactions between atoms is to cause the motions to become mixed up, i.e. to cause order to decrease. As a simple example we consider a box divided into two parts by a partition, such that one side is filled with gas and the other is empty. When we remove the partition this will tell us something important about the initial distribution of positions of the atoms: they will all be on one side. Regardless of their velocities and precise positions at this initial time, we can be sure that at all (sufficiently) later times, there will be essentially equal numbers of atoms on both sides. Thus we observe an apparent loss of order, which distinguishes one time direction from the other. I inserted the word "apparent" in the previous sentence because there will always be an order in the motions of the atoms, namely it will be such that if we reversed their velocities then at some time they would all come to be at one side of the box. However, such an order is not apparent.

The fact that the (apparent) order in the universe is always decreasing is one way of stating the familiar second law of thermodynamics. Although this is sometimes described as a law of physics, and is (rightly) often used as such in many successful applications, it is not a consequence of what we believe to be the basic laws of physics. Rather it is in some sense a consequence of the particular conditions in the world at the present time: we see the world moving to disorder because there is so much order in it. How the order got there in the first place is clearly a very important question, as is the question of whether physics will ever be able to supply the answer.

If we allow ourselves a little excursion into mathematics, then it is easy to give a simple example of a system that is not invariant under time reversal, but which obeys an equation that is. We consider the equation

$$\frac{d^2w}{dt^2} = \lambda^2 w, \qquad\qquad 8.5$$

for which the general solution has the form $w = A\,\exp(\lambda t) + B\,\exp(-\lambda t)$, where A and B are arbitrary constants. Clearly by a suitable choice of "boundary condition", for example by specifying that $\frac{1}{w}\frac{dw}{dt} = -\lambda$ at $t = 0$, we can make A equal to zero, so that the solution becomes

$$w = Be^{-\lambda t}, \qquad\qquad 8.6$$

which decays with time (see figure 8.4), and therefore distinguishes a time direction. If we regard the variable w as representing the amount of order in a system, then equation 8.6 would show the order reducing in time. (It is perhaps not necessary to say that the real world is not so simple!)

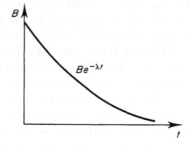

Figure 8.4. A function that "decays" with time, and is therefore not invariant under time reversal, even though it obeys an equation that is, i.e. equation 8.5.

We note in passing that if we use quantum theory to describe the world, and only allow the time evolution given by the Schrödinger equation, then the second law of thermodynamics is not just not obtainable, it is strictly incorrect. In such a theory the amount of order, defined rigorously as the *entropy*, can be shown to be exactly constant in time.

In spite of the uncertain status of the second law of thermodynamics, it has an all-pervading influence on the world around us. It is responsible for the fact that *change and decay in all around I see.* It is also, at least partially, responsible for another time-directed rule that is crucial to our way of thinking about the world, namely, that we regard *cause* as preceding *effect*. Again we are likely to react to such a statement by saying that it is not really a matter of our "thinking"; it is the way the world actually works. Such a reaction is justified: we see something on our television screens because something has happened earlier in the television studio; the earth is going to heat up through the greenhouse effect because of the use of aerosols during the last few years; etc. Note also that the view was built into our understanding of the whole universe when we worried about the lack of causal contact between different parts (see section 4.5); we assumed that the universe *now* is a simple consequence of the universe at *earlier* times, rather than the other way round. However, at the level of basic physics, we can run the whole thing in the other direction. We are here thinking of course within a causal, deterministic, framework, and it is then as true to say that we can deduce what happens at $t < 0$ from knowledge of the system at $t = 0$, as it is to say that we can deduce what happens at $t > 0$ from the same knowledge. Cause and effect go either way!

For those who know about these things the case of electromagnetic radiation is an interesting example here. We normally write the solution of the (t-invariant) wave equation as the so-called *retarded-wave* solution, which is not invariant under time reversal, and which represents waves travelling outward from the source. It is sometimes considered that we have here inserted a time direction into physics, and the choice is justified on "physical grounds". (A book published in 1988 finds it necessary to remark: *it is a physical fact that what happens at later times cannot alter what happens at earlier times,* prior to solving an equation for waves on strings—as if physical facts could have an influence on the solution of equations!) In fact of course we have done no such thing. The use of the retarded wave, plus *initial* conditions, is an exact, unique solution of the mathematical equation. We could have also written this solution in terms of *advanced* waves, travelling in the opposite direction, together with a specification of the *final* conditions.

We can perhaps make the physical situation a little clearer if we think again about the universe drawn on an x, t plot (with y, z understood). Here the universe just *is*; we do not think of one part of the picture as *causing* the other. Of course there are equations relating the various parts of the picture, but no obvious sense in which one piece is the cause of any other. Note that this gives another way of seeing the difference between our perceptions of time and space: although countless books have been written on whether the world is deterministic in the sense of whether the conditions at one time, say $t = 0$, determine everything at all later times, we never worry about whether the world at $x = 0$ explains, or causes, the world at $x > 0$.

In spite of all these things we will persist, correctly, in seeing the earth warming as being a consequence of the use of aerosols, rather than the other way round! One reason for this is that the causal chain appears much simpler when taken in this particular direction. It is comparatively easy to trace the ascent of gasses into the upper atmosphere where they interact with ozone, etc. On the other hand, if we had to trace backwards all the effects which would ultimately cause me to use the aerosol, we would need to follow the motions of the particles of the air that had been heated through friction in my muscles as I moved my arm, also the motion of the particles of the object I was standing on since these provide the momentum that ensures conservation of momentum as I lift the can, etc, etc and etc! Here we see how all this is related to the ideas of the second law of thermodynamics: the ordering of motions of atoms that would cause the aerosol to be used just seems too much of an accident to be the real "cause". A cause is best seen as something simple, whose effects may well spread very far. We can understand one particular effect very easily by following one line; on the other hand to run the process in the opposite direction we would need to follow all the lines. This is illustrated in figure 8.5.

A second reason may have something to do with the way our minds work. Indeed the idea of cause and effect is to some degree anthropomorphic. We believe that *we* can cause things to happen in the future, so we tend to apply the same idea to the world without. It could well be, however, that there are also influences the other way around, i.e. that we regard ourselves as being able to

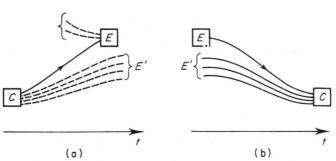

Figure 8.5. Showing how a cause–effect relation is usually simple in that a particular cause (C) produces an effect (E), as in (a), regardless of other influences on E and of the other effects (E'). When we reverse the time, as in (b), there is no simple way in which we can regard E as causing C because the occurrence of C requires also E'.

affect the future but not the past *because* that is the natural pattern we see in the world. Here we are compounding two problems; that of free-will and that of time. If we do this then there is a sense in which the idea of purpose (see section 7.6) might be regarded as inverting the usual order of cause and effect. I want a particular outcome and it is my concept of this "final state" that makes me adjust things so that it is achieved. The future has thus determined the present.

8.4 Memory

We return now to the fact, noted at the start of section 8.2, that we remember the past but not the future. This is one of the most obvious non-time-reversal-invariant experiences. Can we understand its origin?

The first thing to say is that the brain, considered for the moment as a classical physical system, exists at a given time in a single state. This state is connected by the laws of physics to both the past and the future. As long as the brain is isolated, we can calculate its state at an earlier and at a later time, i.o. we can deduce its past and its future. At this stage everything is symmetric in time. As soon as external effects become significant our ability to deduce anything, just from the state of the brain, disappears. To give a simple example, I can see a man walking across the grass outside of my room. This fact affects the present state of my brain, but I cannot deduce from where the man came, or to where he is going. Both the past and the future depend upon factors of which I am not aware. This is illustrated in figure 8.6.

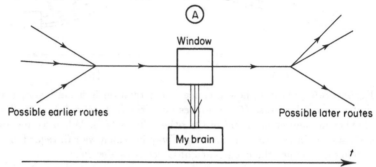

Figure 8.6. Showing how knowledge of a person at A does not allow me to remember from where he came or predict to where he will go.

This does not mean of course that memory of things external to the brain is impossible, indeed we know that it happens. To understand how, we consider a simple form of "memory", namely, a drawer full of photographs. These provide immediate evidence of something that happened earlier because they were produced by a "machine" designed to give a direct correlation between an event (people in unusually new-looking clothes standing by a church door) at $t = t_1$, and a state (a picture of the above people) at times $t > t_1 + T$, where T is the time taken by the photographer to do the processing. This is an example of the sort of process shown in figure 8.5(a). It is not hard to imagine that our brain records images of the past in a similar sort of way.

Have we broken time-reversal invariance? Not really, because the "machine" still operates in a totally t-invariant way. The initial conditions (blank emulsion on the film, state of the camera, etc) were designed with a particular purpose, and it is here that the time direction is inserted. A computer memory provides another simple example of this sort of thing; one particular key will produce a particular state of part of the machine, which has a unique correspondence with the key that was pressed. We should, of course, comment on the use of the word "designed" in the above. This has introduced a concept that lies outside our attempt at a description in purely physical terms. Maybe the second law of thermodynamics, which also seems to require an initial condition helps here (see below).

Clearly it should be possible to design different machines that will do the opposite things, i.e. give us pictures of the future. These would behave as in figure 8.7. In fact, we then call the pictures "blueprints" or "plans" rather than photographs. As we saw in the previous section, when we plan future events in our mind, we are doing the time-reversed process of remembering past events. In this sense "purpose" is the time reverse of memory.

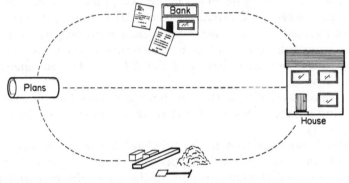

Figure 8.7. Showing how a plan plays the role of a photograph of the future. Regardless of where the bricks, etc, are, it will produce the house.

All this, of course, leaves a big question. It may be that, from the point of view of physics, we can satisfy ourselves that there are equivalent processes in both time directions, but we still have the problem that *we perceive these things as very different.* "Memory" and "purpose" are not different words for the same thing; photographs and blueprints are seen by us to be conceptually different. Why?

I can suggest four, not necessarily mutually exclusive, answers. The first would try to relate the difference to the second law of thermodynamics. The production of a memory state, which at first sight seems to correspond to an increase of order, in fact requires a general decrease in order when everything is taken into account. In other words, the increase in order in the brain is much smaller than the decrease that is required elsewhere in actually producing the required machinery. Thus the direction in which memory operates is correlated with the direction of decreasing order implied by the second law of thermodynamics. There is also a sense in which memory is irreversible; "unmemorising", which we usually call forgetting, can happen, but in a very different way to memorising. Further comments on this idea are given by Hawking (1988).

The second suggestion is the one that we have already mentioned, namely, that it is related to the fact that we believe we can alter the future, whereas we have no corresponding belief about the past. This of course raises many new problems, and it is for example not clear in which direction we should consider the cause–effect relation to operate here: are memory and purpose different because we believe we can change the future, or do we believe we can change the future because of the difference between memory and purpose?

A third possibility is that we have lost something crucial by ignoring quantum theory. We shall return to this possibility in chapter 11.

Finally, we come back to the "obvious" answer to the problem, namely that the past has "happened". This is a strongly held conviction, and is what crucially distinguishes the past and the future in our minds. What does it mean? We study this question in the next section.

8.5   Movement through time

When I was very young, my mother used to warn my brother and me against doing anything we might later regret with Fitzgerald's uncompromising words:

> *The moving finger writes; and, having writ,*
> *Moves on: nor all thy piety nor wit*
> *Shall lure it back to cancel half a line*
> *Nor all they tears wash out a word of it.*

In a letter to a bereaved friend, Albert Einstein wrote:

> *The distinction between the past, the present and the future*
> *is a stubbornly persistent illusion.*

These quotations represent two contrasting views about time. The first is the way in which it is perceived by the conscious mind; the second is the statement of physics.

We regard ourselves as moving through time. As we move we interact with the world, maybe changing it, certainly being changed by it. In other words there is a "now", which is part of total reality, and which alters its position on the time line. Let us try to see what this might mean on our x, t plot, which as we saw earlier is the stage on which the universe is. In figure 8.8(a) we see such a plot. It is supposed to represent **everything**; all that was, that is and that ever shall be. There is, however, something missing from it: there is no indication of "now". We try to put this right in figure 8.8(b), in which now is shown as January, 1990 (i.e. now). In figure 8.8(c) we see a similar plot in which now is in the year 2000. But, which plot is correct? How can I have two different plots, each of which is supposed to be everything that is? Contrast this with what we would say if I had instead labelled the axes x, y. Then it would be in order to say that the second picture would be the first at a later time. We would see that the object called "now" is moving, i.e. changing its position in space as time changes. In our x, t plot however (and recall that this is really supposed to include y and z), we do not have anything that can change to allow "now" to move. Of **what** is its position supposed to be a function?

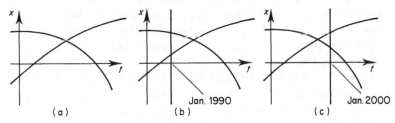

Figure 8.8. (a) is supposed to be a picture of the world; all that was, that is, and ever shall be (for simplicity it is a very simple world, consisting of just two particles moving along an x-axis). (b) and (c) are similar pictures, in which the point "now" has been added. Which is correct?

The only possibility here seems to be to introduce some sort of different "psychological" time, and to say that my position on the time axis varies as a function of my psychological time. It will normally be an increasing function of course, corresponding to the fact that I move forward through time. Then, as psychological

time varies I will have visited an increasing range of times, as shown
in figure 8.9. The times I have visited represent the things that
"have happened"; conversely, "has not happened" means "I have
not been there".

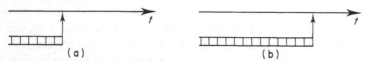

Figure 8.9. Showing how "now" moves along the time axis as psychological time increases. (b) is at a later psychological time than (a).

It is interesting to note that in the "time travel" of science fic-
tion, these two types of time are explicitly assumed. When I travel
backwards in time to watch the Great Wall of China being built,
my *own time* still goes forward. Hence I see the wall being built
after I made the decision to commence my journey in time (figure
8.10(a)). Otherwise I would have travelled back to a time when
I did not exist. On the other hand, when we wish that we could
undo what has happened, to have another chance, we are thinking
of something different, i.e. going back to an earlier point of *our*
time as in figure 8.10(b). Of course, as far as we know, neither of
these processes is actually possible (although some of the so-called
"wormhole" solutions of the equations of general relativity seem as
though they might allow something rather similar). What *is* possible is "memory", which can be thought of as moving our point
of concentration back from "now" to some earlier time, as shown
in figure 8.10(c).

Our sense of moving through time can also be regarded as analogous to reading a book. We only read part of each page, of course,
because we are only aware of a very small part of all that is. This
analogy, however, misses something that is a very important part
of our experience; namely, the experience of free-will that we have
discussed in the previous chapter. We do not only *read* the book,
we help to *write* it! The ideas of time and of free-will are closely
related, and there is no doubt that part of the difficulty in understanding the latter is that we do not understand time.

8.6   Time and space

We have seen that in spite of what we said in section 8.1, there

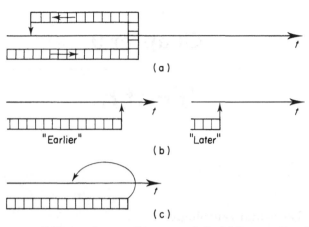

Figure 8.10. Different forms of time travel. In (a) I see earlier times at later psychological times. In (b) the second picture, corresponding to an earlier psychological time, is supposed to happen after the first, which seems to require yet another type of time to understand the "after". In (c) memory is depicted as a movement of concentration away from now.

are several aspects of time that make it very different to any of the components of space: time is directional, it flows, it has a "now", it is associated with purpose, memory and causality in a way that space is not. Maybe the elegance of special relativity has fooled us into a "symmetry" that is only a local illusion. Perhaps, after all, there is a unique, preferred, rest frame in the universe. Certainly some aspects of quantum theory suggest this. Cosmology of course provides such a preferred frame, for example, the frame in which the micro-wave backgound is isotropic, but it is not obvious how or why this should be relevant to the discussion of this chapter (or to quantum theory).

Even more revolutionary (from the point of view of special relativity), is the suggestion that the universe actually exists in space, rather than in space–time, and that "time" is somehow a derived concept. This is suggested by the work of Hawking (1988) and others on the *wavefunction of the universe*, which is supposed to be a complete description of the universe, but which does not depend upon any variable associated with time, i.e. in other words, it is constant. How then can the apparent change which we see in the universe arise? To explain this we need to know some quantum theory, so we will defer further discussion until section 10.7.

Chapter 9

Truth

9.1 Facts and tautologies

There are several ways of interpreting Pontius Pilate's famous question: *What is truth?* Was he enquiring about which of a set of reported stories were correct, or was he asking a more fundamental question about the nature of truth itself? Here it is the larger question: *What does the concept mean?* with which we shall be concerned. We shall see that although we tend to regard *THE TRUTH* as something certain and existing, independently of us, there are aspects of it which may require conscious mind before they can emerge.

We begin by noting that there are obviously many different sorts of "truth". There are statements about things, which can either be true or false. For example, the statement that I am now sitting is true, although there is no reason why it had to be true. Similarly the statement that most trees have green leaves is one that is deduced to be true from observation, unlike the fact that 22 plus 32 is equal to 54. A discussion of truths of this type, which we can refer to as **factual** truths, is given by Carr (1988). It is difficult to give a precise definition of the classification since we can never really know whether a given fact is *necessarily* true or not, and in any case this must depend on what we are allowed to assume. For example, the greenness of trees is a consequence of the chemical structure of chlorophyll, and, given the laws of physics, this could not be different to what it is. Carr defines factual truths as those that are non-mathematical, non-linguistic, and non-logical.

The opposite to factual truths are **tautologies**. These are not necessarily statements *about* anything, but are deductions we can make, using the rules of logic or mathematics, or simply the meaning of words, from other, assumed true, statements. For example, we can suppose that a, b and c represent three ordinary numbers. Then the following statement is true:

$$\text{if } a > b \text{ and if } b > c, \text{ then } a > c.$$

This statement does not depend on anything else for its truth (unlike the statement that I am sitting, which would be false if in fact I am standing). Note in particular that it remains true, though not useful, even if we choose the numbers to be, for example, $1, 2$ and 3.

Another example of a tautology would be the following:

$$\text{if } A \text{ is either } B \text{ or } C$$
$$\text{and if } A \text{ is not } B$$
$$\text{then } A \text{ is } C.$$

If we take A to be *my brother's place of birth*, B to be *Burma* and C to be *China*, then the statement tells me that if my brother's place of birth was either Burma or China, and if it was not Burma, then it was China. This is certainly a true statement. My information would suggest, however, that the initial axiom is not in fact correct.

It could of course be objected that even so-called tautologies are dependent upon something for their truth; for example they depend on the meanings of words, or mathematical symbols. I believe we can meet this objection by noting that what is at issue is the truth of the meaning of the written statement, and this meaning exists regardless of the particular symbols in which it is expressed.

Tautologies are the basic business of logic and of mathematics. They have the great advantage of certainty; we do not have to argue about them, or to rely on uncertain evidence or memory; the fact that the circumference of a circle on a plane is π times the diameter is true even if I have forgotten it, and it is not dependent on anything else. It is therefore an attractive idea that perhaps we could reduce the major decisions of life and society to the level of tautologies, so that decisions could be reached by calculation rather than by argument. Leibniz was one of the philosophers who was seduced by this sort of idea, and, as we saw in section 2.6, Descartes also found it attractive. It is perhaps not surprising that little progress was made.

A slightly less ambitious aim behind attempts to put thinking on a proper logical basis, was to understand the significance of certain statements which appeared to contain an internal inconsistency. We discuss these in the next section.

9.2   Paradoxes

The problem seems to have begun with a writer in Classical Greece, Epimenides, who once wrote:

Cretans always lie.

The credence we would be inclined to give to this statement would depend upon our knowledge of Epimenides. We would become hopelessly perplexed however on learning that he was, himself, a Cretan. To avoid racial overtones we will use a different version of the same paradox and consider the words:

This sentence is not true.

The sentence appears to say something, but clearly it cannot be either true or false (try either assumption). How can such a contradiction arise?

At this stage the theoretical physicist in me wants to say that we should not really bother with paradoxes of this sort. The above sentence clearly conveys no information about anything other than itself, so how can it matter that what it says about itself is paradoxical? It is rather like a particle that interacts with nothing. Why should we worry about it? Nevertheless, logicians, mathematicians and philosophers have worried about it, so we shall follow them.

Apart from its intrinsic lack of interest, there are two other aspects of the above sentence which might be responsible for the confusion it causes. One is that it is "self-referential", i.e. it talks about itself. (In passing we might note that this already makes it potentially relevant to our discussion, because in thinking about consciousness we are necessarily referring to ourselves.) It is trivial to see that we can remove this need for self-reference by replacing the sentence by two sentences:

The sentence on the line below this is not true.
The sentence on the line above this is true.

Clearly, the paradox has not been removed. The two sentences now form a "closed loop", which basically is equivalent to their having the property of being self-referential.

One resolution of the paradox is to make a precise definition of the class of objects about which we are allowed to make statements, regarding truth or falsity, so that the statements themselves are not in that class. Thus self-referential statements are not allowed, and the problem is solved. However, we do not *have* to define things in this way. Gödel (1931) showed that it is possible to put self-referential statements on a proper logical foundation, i.e. to make them allowed statements in the theory, and hence showed that we cannot escape the difficulty by this route. His work shows that the real problem with these type of statements lies with the other key point about these type of statements, namely that they talk about the concept of "truth".

Before we discuss Gödel's work in a little more detail we shall introduce the idea of an axiomatic system.

9.3   Axiomatic systems

An axiomatic system is defined by giving two things: a set of *axioms* and a set of *rules*. The axioms can be regarded as true statements of the theory, and we would expect that they have the form of certain sets of symbols, e.g. sentences in English, or various mathematical expressions. The rules then tell us how we can manipulate these symbols to obtain new statements. Because these are derived from the axioms, they are also "true" statements. We call them *theorems*.

The game of chess provides a very nice, pictorial, example. We can regard it as a formal system in which all positions of the pieces on the board are possible statements. There is only one axiom, namely, the standard starting position of the pieces. We could of course readily write this in symbolic language using any particular chess notation of our choice. The rules are the allowed moves of chess. To express these in the symbolic language is messy, but of course possible. A theorem is then any position which can be reached from the initial starting point (the axiom) using the allowed moves (the rules). Thus, every position in a game of chess is a true statement, or a theorem, of this axiomatic system. We can prove theorems without "thinking"; the process is completely automatic and it is easy to programme a computer to do it.

Of course, not every possible statement is true in this system, i.e. not every possible set of positions can be reached from the initial one. Last week I was playing chess with my wife. I felt that I was in a reasonably strong position; the attack was closing in! There were a few interruptions, telephone calls, liquid refreshment and such like. Suddenly my apparently invincible strength had disappeared. It was not until a few moves later that I realised why: both my wife's bishops were on white squares! Such a statement is *not* a theorem.

This last sentence is readily seen to be a true fact about the logical system we are discussing. However it is interesting to note that our computer, programmed to produce theorems, would never tell us this. It would not, of course, ever produce a statement in which two bishops of one colour were on the white squares, but it would never tell us that this statement is impossible. We could keep the computer running for a very long time and from the fact that we never reached such a statement we might begin to guess, but we would never *know*. The reason is that this fact, though true, is not a theorem, i.e. it is not one of the statements (positions) that can be reached by following the rules. We realise that it is true by thinking about the rules, not by using them. This is why the computer, which is really pretty stupid, will never get there! We have here an example of what is called *meta-mathematics*, a statement that is not inside the mathematical structure being considered, but is a statement about that structure.

The connection with the previous paradoxes now becomes evident; meta-mathematics is rather like self-referential statements. One of Gödel's great contributions was to show how statements of meta-mathematics could be regarded as statements in mathematics. We describe Gödel's theorem in the next section.

9.4   Gödel's theorem

The story here begins with attempts to give the foundations of mathematics a proper logical structure. The idea of regarding mathematics as a set of rules allowing deductions to be made from certain axioms goes back as least as far as Euclid. Developments in formal logic, and the need to extend mathematics to include the concept of infinity, led Whitehead and Russell, in the early years of this century, to attempt a complete treatment of the principles of all mathematics. In 1910 they published the results of their work in the three volume *Principia Mathematica* (Whitehead and Russell, 1910), which must surely be one of the most "unreadable" books ever written!

The problem, however, was still not properly solved. Hilbert pointed out that it was necessary to show that the structure was internally consistent. In other words, we must show that there is no possibility of using the axioms and the rules to derive theorems which are mutually contradictory. As an example within simple arithmetic, we might suppose that we can deduce that two numbers a and b satisfy $a < b$. Then it is necessary to show that there is not a different set of operations that will allow us to conclude $b \leq a$. Note that the consistency problem arises because of the *abstraction*. A physical theory, or a theory *about* something, can be inconsistent only to the extent that it fails to describe the system, because the system itself exists and therefore must be consistent. Roughly speaking the condition of consistency puts a limit on the number of axioms that are allowed; the larger the number, the more likely we are to be able to produce contradictory results.

There is also some sort of lower limit on the number of axioms. We would like them to be complete, in the sense that they should allow us to prove all true statements of the theory. At this stage we might worry about how this can fail to be the case if true statements are defined to be those that are derived from the axioms. We shall see that things are not necessarily so simple. Indeed, we can now state the remarkable result of Gödel:

> *In any finitely describable formal system that is consistent and that is rich enough to contain the rules of arithmetic there are true statements that are not theorems.*

What this means is that there cannot exist a complete logical basis even for all the results of arithmetic, let alone for all truth!

Before we discuss the signficance of this result, we shall outline how it can be proved. Here we should warn the reader that the original proof occupied many pages, and we cannot hope to do full justice to the theorem. (Readers who require further discussion, but without the full details of formal proofs, should consult Nagel and Newman (1959), Hofstadter (1979), Rucker (1987) or Penrose (1989). A modern treatment including all the technicalities is given in Cohen (1987).)

The first step in the argument is to show that all statements can be mapped into numbers, which we refer to as their Gödel numbers. The word "mapped" is here used in an analogous way to its normal English usage; just as a position on a map represents a particular position on the earth's surface, so each number represents a particular statement. There are many ways of achieving this. For example, we can write all statements in English, and convert them to a number in base 27 (= 1–26, for the letters, plus 0, for the space). Thus "I am" becomes the number $(9 \times 27^3) + (0 \times 27^2) + (1 \times 27) + (13)$. Clearly every statement is associated with a particular number. In general there are different ways of making the same statement, but we can make the number unique by choosing the method for which the number is the smallest.

Conversely, any number can be written in a unique way as a series of letters. Some of these will be nonsensical, others will be meaningful statements which are not either theorems or axioms of the theory considered, finally we will have numbers which correspond to statements which are theorems. We denote the statement "x is the Gödel number of a theorem" by $P(x)$ and the statement "x is not the Gödel number of a theorem" by $\sim P(x)$.

Next we consider statements about variables, e.g. "*var* is odd", or "*var* lies between 18 and 39". Again we can associate each such statement with its Gödel number. Thus x might be the Gödel number associated with the particular statement $X(var)$. Now consider the statement $X(x)$. If x corresponds to the first of the above examples this would be the statement "the Gödel number of the statement '*var* is odd' is odd", and similarly for other examples. We denote the Gödel number of the statement $X(x)$ by $[x]$. Thus the operation $[\dots]$ maps certain integers into other integers.

To complete the argument we let $E(var)$ be the statement defining the property $\sim P([var])$, and let e be its Gödel number. Thus we have that $E(e)$ is equivalent to $\sim P([e])$. However, by definition of $\sim$, the latter is equivalent to the statement that $E(e)$ is not a theorem. Hence we have proved that:

$$E(e) \leftrightarrow E(e) \text{ is not a theorem.}$$

We see that, because of the Gödel numbering procedure, the statement that "this statement is not a theorem" is now an allowed statement in the system. Like the original Epimenides paradox, it is self-referential. Everything is now satisfactory because the paradox has disappeared. This has happened because the concept of truth has been replaced by that of theoremhood. However, we can still ask whether the statement $E(e)$ is *true*. Indeed, perhaps surprisingly, we can even answer it! First, it is clear that we can conclude that $E(e)$ is not a theorem. To see this we note that if the contrary is true, i.e. if it is a theorem, then the right-hand side of the above equivalence relation would be false, from which we could deduce that the left-hand side is also false. Thus $E(e)$ would be a theorem and it would be false. However in a consistent system a result cannot be both a theorem and false, so this is impossible. Hence we deduce that $E(e)$ is not a theorem. Thus the right-hand side, and hence the left-hand side, of the equivalence relation is true. Now we have something rather remarkable, we have proved the existence of a statement, $E(e)$, which we know to be true and which we know is not a theorem!

At this stage, readers will feel that they must have been cheated, and have been led to accept things that they will now want to check more carefully. Some will be content to know that many clever people have tried to find a flaw in the proof outlined above, and have failed. Others will need to consult the references given earlier, where they will be able to find all the details.

Gödel's theorem, published in 1931, must have come as a shattering blow to the logicians and mathematicians of his time. It destroyed for ever the hope of putting mathematics on a consistent logical basis. I was almost tempted to write that it was to mathematics something like what quantum theory was to physics, but this would be unreasonable since it has not been as fruitful of further results and progress. The direction of investigations in mathematical logic changed, and the result was expressed in different forms and generalised in various ways (see, for example, Hofstadter, 1979, or Rucker, 1987).

More generally Gödel's result ensured that mathematics would remain a "creative" intellectual activity; not to be exhausted by the routine application of a set of rules. There would be employment for mathematicians, as well as for computers!

Implicit in the last statement is the belief that mathematicians are not just computers, so it naturally leads us to ask about the relevance of all this to our study of conscious mind. We have shown that there are true statements in any theory which a computer can *never* prove, but which *we can see are true*. It appears that conscious minds can learn things about any logical system that a computer, following the rules of the system, can never discover. If we accept this at face value it seems to say that a computer can never model a conscious mind. Such a view was expressed by Lucas (1961): *Gödel's theorem seems to me to prove that mechanism is false, that is, that minds cannot be explained as machines.* Attempts to rebut this claim are given in Whitely (1962), Good (1967, 1969), Webb (1968) and in Hofstadter (1979). It is defended in Lucas (1967) and, more recently, in *The Emporer's New Mind* by Penrose (1989).

On reading these arguments, it seems to me that again everything depends on what we *mean* by words like "machine". Lucas regards a machine as something that can only follow rules and reach theorems; it cannot ever "see" a result. It cannot sit back and survey the problem, and realise that something is true. On the other hand *I* can. But why? The reason of course is that I am conscious. (No, that is not necessarily the *reason*, rather I should say that it is associated with the fact that I am conscious.) So Lucas, in saying that a machine cannot do these things, is essentially *asserting* that a machine is not conscious; he is not *proving* it. (I would guess that this is an area where "proofs" are unlikely ever to be available; certainly not until we know how to define the words better!) We are back with the question of why I cannot put whatever it is about me, that makes me able to go beyond the algorithms, into a "machine", a question which is surely related to the problem of why I cannot put whatever makes me conscious into a machine.

Hofstadter tries to argue against Lucas by claiming that we are really as limited as the computer. He shows that any logical system can always be extended, so that the true statements that are not theorems of the original system become theorems of the new system. (Of course when we have done this we know from Gödel's theorem applied to the new system that there will be some new true statements that are not theorems.) He then claims that we are really operating as computers in the extended system, and that this is what gives us a power that the computer does not have. Once we have discovered a truth we could programme a computer to obtain it as a theorem, and this is a process that we could continue indefinitely. Ultimately, he claims, everything will become so complicated that we will not be able to see any new results that the computer cannot prove. I have probably not done justice to the argument but it seems to me that it misses the point, which is not so much that I will always be able to argue in a way that transcends the rules, but that I can *sometimes* argue this way. We *have* a concept of truth, which the computer has not got, at least not unless, or until, it is conscious. There is something about truth, or perhaps rather I should say about the recognition of truth, that requires conscious mind for its existence; like free-will, or red, or beauty, it is one of the properties of consciousness.

Tipler (1989), in a review of the book by Penrose mentioned above, makes similar arguments to Hofstadter. He objects to the claim of Penrose that meaning requires people, but in my view the objection is seriously weakened when he is led to make statements like: "The meaning in the programmes being run inside human brains comes in part from our evolutionary history—programmes coded in our ancestors' DNA were varied at random, and the meaningful programmes were preserved ... (this is what is *meant* by meaningful) ...". Is it?

We use anthropomorphic terms to describe our computers, which is why we become annoyed when they make a mess of our work by "doing stupid things" (because we have issued a wrong command). They should have "known" that we could not possibly have meant that! The problem is computers do not "know" anything:

> ... the computer does not have a way of judging truth; it is only following rules. It does not 'see' the validity of the Gödel argument. It does not 'see' *anything* unless it is conscious! It seems to me that in order to appreciate the validity of the Gödel procedure—or, indeed, to *see* the validity of *any* mathematical procedure—one must be conscious. One can follow rules without being conscious, but how can one *know* that those rules are legitimate rules to follow without being, at least at some stage, conscious of their meaning? (Penrose, 1987).

Later (section 11.7), we shall see how quantum theory also suggests that the idea of "knowing" is somehow inappropriate for physical systems.

Gödel solved the problem of the Epimenides class of paradoxes by recognising that there is a crucial distinction between true statements and theorems. An automatic, *algorithmic*, procedure, which is all that a computer is capable of, can only produce theorems. Somehow, "truth" is a bigger thing; the canvass on which statements are true or false is a larger one than that on which they are theorems or non-theorems.

In this section we have been concerned with a result of rigorous mathematics. It is perhaps not surprising that the result concerns something that might be regarded as a trivial and unimportant issue, i.e. the status of a statement that says nothing about anything other than itself. However, there is a very clear link with something potentially much more important; something which we have already met:

9.5   Does a machine tell the truth?

We recall that in section 3.3 we noted that we would not be convinced if a computer printed out a message: *I am conscious.* This would tell us that the design of the machine, and the programme that had been fed into it, was such as to cause this message to appear. It would tell us nothing about the mental state of the machine. This is not because a computer has a desire to lie. It just has no idea what words mean, and has no concept of truth or falsity. It simply prints out the message that it is told to print out!

Now I ask myself an important question. Will I *ever* be able to take seriously a message like that above coming from a computer? Or, to put this another way, what properties will my computer have to have before I will regard its answer to the question: *Are you conscious?* as having any relevance whatever to the issue? I believe that this is a question which anybody who believes that he can make a conscious computer must face. The problem with it is that I cannot see that there is any satisfactory resolution. I know that the answer that I get out will always be determined by the programme in the computer. Of course I may not be able to predict what it will be, but that is clearly irrelevant. For example, it may be that the machine is programmed so that if the answer to a very complicated calculation is greater than 40 then it will answer "yes", and otherwise will answer "no". The answer will tell me about the calculation, but it will tell me nothing about consciousness!

Similarly, I could build a random number generator into my computer, so that the answer to my question is *not* determined by the programme. The answer would then be different every day, which would at least provide a bit of variety. It would not, however, help us to get a *true* answer to our question. There *is* an answer, and it has nothing to do with a random number! At this stage I am tempted to write that I know that I will never be able to build a computer that will tell me, convincingly, that it is conscious, and hence that I will never be able to build a conscious computer. Three things deter me however. The first is that "never" results usually show nothing other than lack of imagination. The second is that I would be less confident if I could build a genuinely *quantum* computer, whose output would depend upon microscopic quantum events, where the inherent uncertainties and non-localities of quantum theory would become relevant. In one sense this would be the same as saying that the machine would then be "open" to influences from outside its mechanical parts (*more* than just the quarks and leptons which apparently make it up). We shall discuss this further in later chapters.

The other hesitation concerns the fact I do believe *myself*, and similarly, I believe *you*, when you make such a statement, and I do not really know why! If I did know why, then I would perhaps know enough to solve the problem of how to make a conscious computer.

Of course, I can make a try by saying that I believe *you* because you are rather like me. Then I have to ask why I believe myself. Here the answer would seem to be that I know the statement corresponds to an experience that I actually have. But have I any right to say this? Suppose, for the moment, that I accept a physicalist and deterministic doctrine. Then, when I say: *"I am conscious"*, the statement can have no more significance than when a computer says it. The statement is made not because it is true, but because the particles in my brain move in such a way as to make me say these words. It is just a consequence of the initial conditions at the start of the universe, or, if you like, of the way I was made and programmed.

Arguments such as this have often been used to demonstrate that physicalist doctrines must be wrong. For example, Swinburne (1987) writes:

Epiphenomenalism is self-defeating; if it were true we would never have any justification for believing it to be true. The epiphenomenalist must hold that our judgements (our conscious expressions of belief to ourselves) are caused by brain states and these brain states by other brain states, and these ultimately by other physical states outside the body. They will not be formed by chains of cogent thoughts. Now, the belief that my belief B is formed through a causal chain as such in no way impugns my justification for holding B. Thus my belief B that there is a table in front of me is caused by a series of brain and then bodily and then extra-bodily causes in no way impugns my justification for holding B, since I will believe that among the causes of B is the table in front of me; and so, that I would not have held B but for B being true. But the belief that there are no items among the causes of B puts B into a certain category, the category of perceptual or semi-perceptual beliefs which are non-reasoned responses to the environment—beliefs which are not held because they are justified by other beliefs. Such beliefs form the foundations on which reasoned beliefs are built—or so we normally think. But if epiphenomenalism is true, it will not be so; *all* beliefs—be they about cosmology, or quantum physics, logic ... or epiphenomenalism itself—will be grounded in the same category of intuitive responses to the environment, not grounded in other beliefs.

Clearly what is said here about epiphenomenalism (see section 5.4) is equally true of any totally physicalist or deterministic doctrine. The debate about it goes back to Epicuras. For further discussion we refer to Popper and Eccles (1977, p.75) and especially to the careful treatment in chapter 6 of Honderich (1988) where many further references are given. In spite of the fact that Honderich finds weaknesses in the argument, it is very persuasive. Indeed the problem with it is that it seems almost too powerful! I have to ask again what is it that convinces me that I should (and do) take notice of my judgements; I do believe that they are based on reasoned arguments. In particular, I do believe that when I say things about myself, the statements are correlated with what is actually the case. The fact that I can, deliberately, lie, is of course further confirmation of the same point, since then I *know* that I am lying. In some sense of course this sort of correlation between facts, and statements about facts, is very familiar in the world. As an example, every Sunday, sheets of paper appear on which there are particular patterns of black ink which correlate exactly with the passage of plastic balls over certain lines, at certain times, on the previous day. Now I have no reason to believe that the atoms that give rise to these printed football results have ever failed to obey the laws of physics, but at the same time I would find it hard to explain what is causing the correlation if I were not able to use the idea of conscious mind (cf. the discussion in section 7.5).

Hawking (1988, p.12) raises a similar point when he notes that a "theory of everything" would have to explain the person assessing it, so it would determine whether he believed it or did not believe it. What credence then could we give to either belief? He suggests that natural selection might favour systems that have beliefs that correlate with the true facts, and that this might explain how conscious systems acquire a notion of truth. If this is so, then the answer to the problem of when I would believe a computer that told me it was conscious, would be after many millions of years of development, with many possibilities abandoned and much fierce competition, etc. But would I "believe" it, or would I rather conclude that it had given me the answer that it thought would be the most likely one to make me want to keep it? Since, ultimately, such a method of selection depends on the laws of physics, it is clear that the answer given is dependent upon those laws. It is a pleasant thought, and a good note on which to end this chapter, that the pursuit of truth might be somehow a property of the laws of physics!

Chapter 10

Quantum theory

The purpose of this chapter is to give an elementary introduction to quantum theory, concentrating in particular on those aspects of the theory which seem as though they might be relevant to the topic of this book. We shall use very little mathematics in the main text. Readers who are not put off by a few equations will be able to obtain more details of some points in the figure captions. I have written an introduction to quantum theory at a similar level, though at greater length, in *The Mystery of the Quantum World* (Squires, 1986). Other books with essentially the same aim are d'Espagnat (1983), Polkinghorne (1984), Gribbin (1984) and Rae (1986). For those who wish to make a proper study of quantum theory there are many books available. Most of these do not dwell greatly on the interpretation problems, but Sudbery (1986) is a recent exception. The collection of papers by John Bell (Bell, 1987a) is essential reading for all serious students of the foundations of quantum theory.

10.1 Particle/wave duality

In section 4.3 we described how quantum theory revolutionised physics and initiated the marvellous progress of this century. We also suggested that a price was paid for this success: crudely speaking, we have no idea what is going on! In particular we saw that those things which we tend to think of as particles, e.g. electrons, neutrons, etc, interfere in exactly the same way as was first observed much earlier for light, and which told us that light was a wave phenomenon.

In figure 10.1 we show the results of some recent very beautiful interference experiments using slow neutrons. The neutron beam is separated into two parts which follow different paths to the detector. Contrary to the situation described in the interference experiments of section 4.2 (figure 4.4), the path lengths are the same, so the two waves will be identical and will reinforce each other. However, it is possible to introduce something into one of the paths which gives the effect of changing the path length, and hence introduces the possibility of destructive interference. The rate at which neutrons arrive at a fixed detector should therefore vary with the amount of the change introduced. As the figure clearly shows, such interference does indeed occur, in perfect agreement with the predictions of careful calculations of the expected interference pattern. The caption to the figure gives a few more details, and we refer to Rauch (1984) for further discussion of this type of experiment.

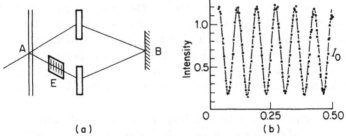

(a)   (b)

Figure 10.1. (a) The neutron interference experiment. The beam of neutrons is split into two at A and brought together again at B. The material at E introduces a so-called "phase shift", which is effectively a change of path length. (b) Plot of counting rate against phase shift. The points are the experimental results and clearly show interference effects. The line is the prediction of quantum theory, as explained in section 10.3, and we see that there is perfect agreement.

We shall now remind ourselves that this behaviour is not compatible with the picture of neutrons as ordinary *(classical)* particles. To see this we imagine that one path, say the lower one, is blocked. Then neutrons travelling along the top path will give a particular intensity at the detector, which we call N_T. Now we block instead the other path, so that particles can only travel along the lower path. These give an intensity which we call N_L. The intensity when we allow particles to travel along both paths should then be the sum of these, i.e. $N_L + N_T$. The reader should pause here to be convinced that this result cannot possibly be wrong! The only conceivable doubt that might arise is if the neutrons somehow "get in each other's way". However, such a doubt can be removed if it is realised that in the actual experiments the intensity of the initial neutron beam is such that the neutrons actually pass through one at a time; there are seldom two neutrons in the apparatus together. Since both N_L and N_T are positive, it "follows" that the intensity when both routes are open will always exceed that when either is closed. This is not what is observed.

It is clear then that, at least in some circumstances, objects that we once thought to be particles are in fact waves, on the same criteria that are used to identify light as a wave phenomenon (section 4.1). If we try to escape from this problem by saying that we were mistaken in believing that the objects were particles, then we have to ask why they make tracks in photographic emulsions or flashes on screens, why they satisfy the momentum conservation equations of colliding particles, etc. As we have seen, the confusion is completed by the fact that light itself is also known to have a particle-like behaviour, the particles being called photons. The history of physics might well have been very different if this property of light, rather than the wave property, had been discovered first.

It is worth noting the slightly ironic fact that, given the full formalism of quantum theory (which did not exist in 1905) to describe the atoms (as in section 4.3), the results of experiments on the photo-electric effect can be explained without explicitly introducing photons. It is only in recent work of Aspect and Grangier (1987) that direct experimental evidence for photons has been obtained. Details of their experiment are given in figure 10.2, where we also note that they used the same source of photons to demonstrate interference; thus revealing in one experiment the full mystery of the particle/wave dilemma.

The apparent contradictions that we have met in this chapter, and earlier in section 4.3, concern the results of experiments. We shall see how quantum theory provides a formalism which allows us to calculate these results, and also those of many other experiments. It does not, however, resolve the contradictions. They are still present, and are usually referred to as the "interpretation" or "measurement" problems of quantum theory. We should remember that they are not, primarily, difficulties of a particular theory,

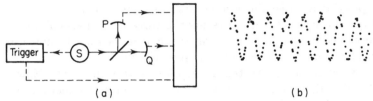

Figure 10.2. (a) The source emits two photons, the first of which is used to switch on the apparatus. It is observed that the other photon always goes to one, not both, counters (P,Q). In contrast, a wave would be expected to excite both, since a part would be transmitted and a part reflected. (b) Here the interference pattern which is built up when the detectors are replaced by mirrors is shown. This is what would be expected for waves.

but of experiments. Many years ago Einstein wrote *the more success the quantum theory has the sillier it looks* (see Pais, 1982, p.399). The success has continued, and if it is still silly, which, by the standards to which we are accustomed in science, it is, then this is because the world is silly! It is this quantum theory which we shall describe in the present chapter, leaving discussion of its possible interpretations to chapter 11.

10.2 The wavefunction of quantum theory

Quantum theory was developed as a response to the above facts, and to others of a similar nature. Although initially it began as a few tentative ideas, it has become an elaborate, and enormously comprehensive, theory with successful applications far beyond the original experiments it was designed to explain. Here we shall describe some of its key features, and see how they confront the apparently insuperable experimental contradictions that we met in the previous section.

We begin by thinking of a system containing, for example, one electron. If we were describing this system in classical physics we would have to say where the electron is at a given time. Thus, if we label positions by the variable x, we would have to give a specific value for this variable. In general the particle would move so that its position would depend on time. Thus we would denote its trajectory by $x(t)$, with t labelling the time. Note that x labels a position in three-dimensional space, so it is in fact a *vector* quantity (in other words it represents the three coordinates of position discussed in section 8.1). However this fact need not trouble those readers who do not know about such things.

In quantum theory the description is very different. Now we describe the electron by a *wavefunction*. This is something that has a value at all points of space. We usually denote it by a Greek letter, e.g. ψ. When we wish to remind ourselves that the value depends on the point of space we are considering, and upon time, we write it as $\psi(\mathbf{x}, t)$. The state of our system at a given time is specified by the values of ψ at all points of space. These values can be either positive or negative (or of course zero) so, as we require, there will be the possibility of interference. (In general ψ is a so-called "complex number", but this is a complication we can ignore for the purpose of our discussion.)

Knowing how a system is described at a given time does not say anything about the dynamics of the system, i.e. about how it changes with time. For a classical system the dynamics is contained in Newton's laws of motion, which tell us that the particle is subject to an acceleration due to the forces that act on it. The appropriate equation of motion is equation 8.1. In quantum theory Newton's laws are replaced by the Schrödinger equation, which was also quoted earlier (equation 8.2). This equation, which of course contains terms corresponding to the forces, gives the time evolution of the wavefunction. Thus, if we know ψ at some time t_0, and at all $\mathbf{x}$, we can use the Schrödinger equation to calculate it for all future times. An example of a wavefunction, and of its possible evolution in time, are given in figure 10.3.

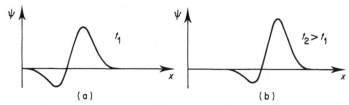

Figure 10.3. Showing a possible wavefunction for a particle moving along a line (i.e. in one space dimension) at two times, t_1 (a) and $t_2 > t_1$ (b). In this particular state the "average" particle position (see section 10.3) has moved to the right.

At this stage readers may well say that this is all very well as a description of a wave, but we are supposed to be describing an electron, which we believe to be, in some sense at least, a particle. When we know the wavefunction, where, for example, is the electron? We discuss this in the next section.

10.3   The probability interpretation of the wavefunction

In orthodox quantum theory (alternatives are discussed in section 11.5), the wavefunction is the *complete* description of the electron; thus reality does not contain a "position" for the electron. This is an illustration of a very general fact: quantum theory itself is not troubled by particle/wave duality because it does not *have* any objects corresponding to classical particles. On the other hand it does have waves, and, as we have seen, the calculations that we make in quantum theory are concerned with a wavefunction which obeys a *classical, deterministic,* wave equation.

The fact that we can, and do, "observe" or "measure" positions of particles is met by adding to the laws of quantum theory an extra rule. It is this rule, or perhaps we should say the need for it, that causes all the interpretation problems. The rule says that the chance of finding the particle at a given point is proportional to the square of the wavefunction at that point. More precisely:

when I measure the position of the electron I will find a result $\mathbf{x}$*, with probability proportional to the value of* $|\psi|^2$ *at that particular position* $\mathbf{x}$*.*

We note that this value, as required for a probability, is always positive. What the rule means is that the particle is more likely to be where the magnitude of the wavefunction is large than where it is small. This is illustrated in figure 10.4. Actually, since position is a continuous variable, the probability of getting *exactly* a particular value is vanishingly small, so we really should define the probability a little more carefully. This is done in the caption to figure 10.4 for the sake of interested readers.

In order to see how interference effects can occur, we suppose that there are two contributions to the wavefunction in some region of space. We call these ψ_1 and ψ_2. They might, for example, be contributions from two different paths in some sort of two-slit

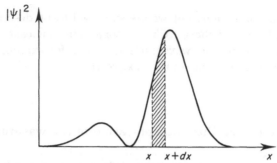

Figure 10.4. Showing $|\psi|^2$ for the real wavefunction given in figure 10.3(a). The probability of finding the particle between x and $x + dx$ is proportional to the shaded area, and is given by $|\psi|^2 dx$, provided the wavefunction is "normalised" so that the particle is certainly somewhere, i.e. $\int |\psi|^2 dx = 1$, the integral being over the whole line.

experiment. The total wavefunction is then given by

$$\psi = \psi_1 + \psi_2 \qquad\qquad 10.1$$

and hence the probability distribution by

$$|\psi|^2 = |\psi_1 + \psi_2|^2, \qquad\qquad 10.2$$

which of course is **not**, in general, equal to $|\psi_1|^2 + |\psi_2|^2$. For example, if $\psi_1 = 3$ and $\psi_2 = -2$, then $|\psi_1|^2 + |\psi_2|^2 = 9 + 4 = 13$, whereas $|\psi_1 + \psi_2|^2 = 1^2 = 1$. A more realistic example is that of figure 10.1. If the two beams are ψ_1 and ψ_2 then at B we will have $\psi_2 = \exp(i\chi)\psi_1$, where χ is the phase shift. Hence the counting rate is $|\psi_1 + \psi_2|^2 = |\psi_1|^2|1 + \exp(i\chi)|^2 = 2|\psi_1|^2(1 + \cos\chi)$. We note here the very important fact that in quantum theory we add *wavefunctions* and not just probabilities, which are the squares of the magnitudes of wavefunctions and are therefore always positive. This distinguishes probability in quantum theory from normal, classical, probability where interference does not happen.

It is clear that the range of possible values that can result from a measurement of position depends on the shape of the wavefunction. If this is sharply peaked about one position, as for example in figure 10.5(a), then only a narrow range of possible values is likely. On the other hand, if the wavefunction is more like that of figure 10.5(b), then there will be a large uncertainty in the predicted result of a position measurement.

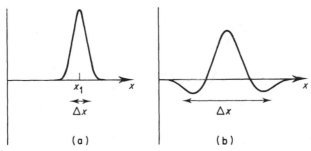

Figure 10.5. A wavefunction for a particle which is very likely to be found close to x_1. The uncertainty is proportional to the width as shown. (b) Here we see a wavefunction for a particle in which there is a large position uncertainty.

It is possible to measure or observe other quantities apart from position. For example we can measure the "momentum" (which is defined in classical mechanics to be the velocity multiplied by the mass). The wavefunction again tells us the probability of getting any particular result for this measurement. The precise formula used to extract this probability from the wavefunction is more complicated than for the position and is explained, for readers with the necessary background mathematics, in the caption to figure 10.6. It turns out that those wavefunctions for which the position uncertainty is small, e.g. like figure 10.5(a), give a large uncertainty in momentum (figure 10.6(a)), whereas those for which the momentum uncertainty is small, as in figure 10.6(b), have a large degree of position uncertainty (figure 10.5(b)). This balance between knowledge of position and momentum is called the **Heisenberg Uncertainty Principle**. It can be derived as an exact inequality:

$$Position\ uncertainty \times Momentum\ uncertainty \geq \hbar, \qquad 10.3$$

where $\hbar$ is the usual notation for $h/(2\pi)$, h being Planck's constant of section 4.3.

We have only learned the very beginnings of quantum theory. In order to understand its many successful applications we would need to develop these ideas much further. However our aim is different, and we are already in a position to appreciate the three ways in which quantum theory has dramatically altered our understanding of the nature of reality. These are the topics of the following sections.

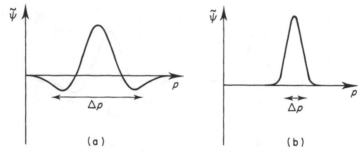

Figure 10.6. Showing the real part of the function

$$\tilde{\psi}(p) = \frac{1}{\sqrt{2\pi}} \int \exp(ipx/\hbar)\psi(x)dx$$

for the two wavefunctions shown in figure 10.5. Since $|\tilde{\psi}(p)|^2$ determines the distribution of momenta in the same way that $|\psi(x)|^2$ determines that of position, it is clear that (a) has a large spread of momenta and (b) a small spread. Comparing with figure 10.5 we see that this illustrates the uncertainty relation (equation 10.3). The average momentum is $\int p|\tilde{\psi}(p)|^2 dp$, which can be shown to be equal to $\int \psi(x)^*(-i\hbar\frac{\partial}{\partial x}\psi)dx$. Thus we can represent the momentum, p, as an operator $-i\hbar\frac{\partial}{\partial x}$. In fact all measurable quantities in quantum theory correspond to operators. A state which has a unique value (probability one) for an observable is in an "eigenstate" of the corresponding operator.

10.4   Quantum theory and determinism

As we have seen, the wavefunction at a given time determines it uniquely at all future times (and indeed at all past times). Hence, at this level, quantum theory, like classical physics, is deterministic. However, we do not measure (observe) wavefunctions, and for the quantities that we do measure, e.g. the position of an electron, the wavefunction does not in general tell us the result we will get; rather it allows many possible results; and it only tells us the *probability* of getting a given value. Thus, if it is true, as orthodox quantum theory says, that the wavefunction is the complete description of the system being considered, then the result of an experiment is never exactly predictable.

Two points should be noted about the above statement. Firstly, the lack of predictability has nothing whatever to do with the difficulties of doing exact calculations, or of making the exact measurements that might be required to know the wavefunction. This point should be stressed because it has often been pointed out that, in practice, the predictability of classical physics is very limited because we really cannot do precise calculations or make arbitrarily accurate measurements. These things are true, but the unpredictability of quantum theory is of a totally different nature; it is a new phenomenon not found in classical physics; it is not about how clever we are at performing experiments or doing calculations, or about how well we need to know the initial conditions, but is about the real world, the world that is out-there, regardless of how much we know about it.

Secondly, the statement that an exact prediction can never be made is a little too strong because the system might be in a special wavefunction for which one particular experimental result is exactly knowable, e.g. if the electron wavefunction is zero at all points except one, then quantum theory tells us that the electron will be at that point. The particle is then said to be in an "eigenstate" of position (see the caption to figure 10.6). In general systems will not be in the eigenstates of the things we wish to observe.

The example of passage of a particle through a barrier gives a good illustration of this lack of determinism. In classical physics, whether or not a particle passes through, e.g. whether it can get over the hill, is determined uniquely from the properties of the barrier and the particle. However, in quantum theory, all we have is the behaviour of the wavefunction. This might, for example, be as shown in figure 10.7. Then we can only discover whether or not a given particle has gone through by making a measurement. If we do this many times we will find that the ratio of the number of times that the particle has gone through, to the number that it has not, is equal to the ratio of the areas under the bumps in the wavefunction on the two sides. Thus quantum theory correctly predicts the *probability* that a particle will pass through the barrier. It does not tell us whether or not a *given* particle will pass through.

It is worth emphasising what we have learned in this section. If orthodox quantum theory is correct then *the complete theory cannot in general, even in principle, tell us the result of a simple experiment.* Exciting though this discovery is, it is not by any means the most important of the mysterious and revolutionary

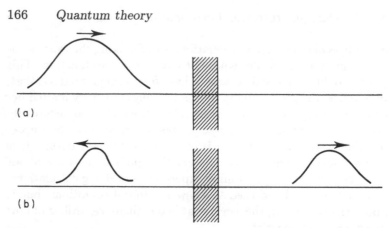

Figure 10.7. Showing how a wavefunction is affected by a barrier. In (a) we see the wave packet approaching the barrier. In (b) we see the situation at a later time; part of the wave has been reflected and part transmitted. The integral of $|\psi|^2$ over the left-hand piece, for example, gives the reflection probability.

features of quantum theory.

10.5   Quantum theory and external reality

Quantum theory does not tell us what *is*, rather it tells us about what will happen when we *observe* (in general it only tells us this in a probabilistic way, as we saw in the last section). To appreciate what this means we compare the situation in classical physics. There an electron is described by a position $\mathbf{x}(t)$. This states that, at time t, the particle is at $\mathbf{x}$, and hence that we will observe it to be at that point. On the contrary, quantum theory describes the particle by a wavefunction; until an observation is made *there is no position*.

It is important to realise that we are not saying here simply that we do not know the position until we measure it. This of course is often the situation in classical physics: the particle is at a particular point but we do not know which. The distinction is crucial for the quantum theory understanding of the double-slit experiment (for example). If we had to say that the electron really followed one trajectory, then there would be no way to explain interference, regardless of whether we know the route followed.

In other words, classical probability theory just deals with *our knowledge* of a system. This is why interference does not happen. On the other hand, the quantum mechanical wavefunction is a real quantity, existing in the external world. Indeed in orthodox quantum theory the wavefunction *is* the external world.

There are two questions that now arise. We have not said what we mean by an "observation", and we have not said in what way an observation affects the physical reality, i.e. the wavefunction. It appears from the above paragraph that observations somehow create positions, but what are the key features of an apparatus that is able to do this, and what does the wavefunction look like after it has happened? If we refer again to the case of a particle coming to a barrier (figure 10.7), these questions can be expressed as: *What do we have to do to the system in order to force a decision as to passing through or not, and what has happened to the wavefunction afterwards, in particular, does it somehow change to a single peak, on one side or the other?* The latter phenomenon is an example of what is called "collapse", or "reduction", of the wavefunction. We discuss it in more detail in section 11.2, where we meet the controversy as to whether or not it actually occurs.

At this stage, readers may well be thinking that, since we have a theory that is supposed to tell us what happens, we should be able to answer the above questions. In one sense they are right, and the problem is that the answer is not acceptable. As long as all systems are described by quantum theory, then observation, of the type described in the previous paragraph, is not possible. This is a startling conclusion, but it is not hard to see why it is so. We have already said that the time dependence of a state in quantum theory is given by the *deterministic* Schrödinger equation. From such an equation we can never obtain the *probabilistic*, i.e. non-deterministic, feature of observation. *Something else is needed.* As we shall see in the next chapter, the various "interpretations" of quantum theory differ in what they use for this something else.

10.6 Quantum theory and locality

The third new feature of quantum theory challenges a very basic assumption that lies at the heart of the scientific way of thinking, and which, roughly speaking, says that we can study what happens in some small region of space, over a small time, without having to worry about what is happening in regions of space very far away. Some sort of idea like this is necessary for any kind of scientific method to work, because we rely on being able to repeat experiments under identical conditions. Of course this is never possible, so we have to be satisfied by an attempt to make the *relevant* conditions identical. There is some hope of doing this provided that it only involves the *local* conditions. For example, if we want to measure the thermal capacity of water we would need to control the ambient temperature and perhaps even the atmospheric pressure on our apparatus, but we would hope that the position of the moons of Jupiter did not make too much difference to our results.

There are two ways in which quantum theory upsets this assumption of locality. The first can be seen immediately from the fact that a measurement to discover whether we have a particle at some position x, on a line say, as in figure 10.3, might give the answer yes. In this case we know that the particle is not at any other position on the line, so something has "happened" (instantaneously?) at all other points on the line. As an example of this we might again consider the barrier experiment of figure 10.7. If we find that the particle has passed through then immediately we know that it has not been reflected, hence the peak in the wavefunction corresponding to the particle having been reflected must either disappear (this is collapse of the wavefunction), or for some reason it must no longer give the probability for the particle to be found in that region (in which case one of the rules of quantum theory is violated).

The other origin of non-locality occurs when we have systems containing more than one particle. To keep things simple we consider a system with just two particles. The wavefunction describing such a system is then a function of *two* position variables, i.e. it has the form $\psi(\mathbf{x}_1, \mathbf{x}_2, t)$. Now this is a very peculiar object. It does not describe the distribution of "something" over all of space because we cannot say how big it is, or, if you wish, how much of ψ there is, at a particular point. We can only give a value to ψ when we give two positions in space or, in other words, when we specify a position in a "space" of six dimensions. An incidental consequence of this is that I am suspicious of attempts to define "physical" things as things that have location in space (see, for example, Honderich, 1988). I tend to believe that wavefunctions are physical things, but they clearly do not just refer to local conditions.

The practical significance of the wavefunction depending on two positions is that the probability of finding the first particle at a given position depends upon the position of the second particle, so, when we measure the second position we effectively alter the probability distribution of the first, even if this is very far away. When this happens the wavefunction is showing the effect of *correlations* in the positions of the two particles. (The word *entanglement* is sometimes used in this context.) It is the presence of such correlations that is the basis of **Bell's theorem** (Bell, 1964), a much discussed result which has beautifully demonstrated the non-locality of the quantum world.

In order to explain Bell's theorem, we introduce a property of a particle which can take only two values. This is much simpler than position which has an infinite number of possible values. The obvious example of such a property is "spin", which for an electron, when measured in a particular direction in space, can take values + or $-1/2$ (in suitable units). The nature of the property is not important for our discussion, but we will continue to refer to it as spin, and call the two values simply + and −, respectively. When we measure consecutively the spins of two particles there are four possible results: $(+, +), (+, -), (-, +)$ and $(-, -)$. The wavefunction will tell us the probability of getting any of these four. If the probability of getting + for the first particle is independent of what we get for the second, then the spins are said to be *uncorrelated*. In general, however, this will not be true. An extreme case of correlation occurs when the particles are in what is called a state with zero total spin. Then the wavefunction allows only two possible results for the spin of the two particles, namely, $(+, -)$ and $(-, +)$. There is now perfect correlation in the spins; if one is positive, the other is negative, and *vice versa*; there is no chance of finding both to be positive or both to be negative.

It is this situation that is utilised in the modern version (due to Bohm) of the, so-called, Einstein–Podolsky–Rosen (EPR) paradox. If we have two electrons in a state of zero total spin, and far apart, then when we measure the spin along a particular direction of one of them, the spin of the other, in the *same* direction, immediately becomes determined, even though the particle itself apparently has not been affected. It was argued that this meant that in fact the spin must have been a part of physical reality even before the measurement. However, this is not possible, because we could repeat the argument with spin measured along *any* direction, thereby concluding that the particle had a definite spin value along any direction. This would contradict a crucial result of quantum theory, namely, that a state cannot have precise spin in all directions. (This is rather like the fact that it cannot have both precise momentum and precise position.) Figure 10.8 and the associated caption give a few more details.

The EPR problem as outlined above requires us to *think about* the possibility of doing two different experiments, i.e. of measuring the spin of particle 1 in two different directions. We cannot of course actually *do* both the experiments; if we decide to do one of them then we will inevitably spoil the correlation between the two particles, and no further information about particle 2 will be obtained if we subsequently do the second measurement. This means that there is no obvious conflict between locality and experiment; the conflict is between locality and a particular understanding of what a complete description of reality should contain.

All this changed in 1964 when John Bell showed that it is possible to use the locality assumption to obtain restrictions (the "Bell Inequalities") on the results of a series of measurements in an EPR type of experiment. These inequalities refer to averages over many repetitions of the experiment, and in this way they get over the fact that we cannot do spin measurements in different directions on one particle. Bell also showed that the predictions of quantum theory violated these restrictions.

Readers who do not wish to learn the details of the Bell inequalities, or how they are proved, should now jump to below equation 10.5. For others, we again consider a situation where two particles, in a zero total spin state, move apart as in figure 10.8. The locality assumption is that, after a certain time (e.g. when the particles are many miles apart!), they are independent of each other. In other words each particle carries with it all possible information that will determine, as much as this can be determined, its future behaviour. Note that we have worded this in such a way that it allows for the possibility that there is some random element in the behaviour of the particles. In this case the information that the particle is supposed to carry with it is the rule that determines the probability of obtaining a particular behaviour.

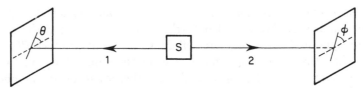

Figure 10.8. The Bohm version of the EPR experiment. Two particles move away from S in a state of total spin zero, which means that if one is + along a given line, the other is − along a parallel line, i.e. the spins point in opposite directions. Thus if we measure the spin of **1** along the direction θ and find +, then **2** will have spin − along the direction $\phi = \theta$. Thus, *without doing anything to this particle*, we can find its spin in any direction. According to EPR this means that it must have a definite spin in all directions. This would violate the spin analogue of the uncertainty principle. In the Bell inequality, discussed in the text, we use two values of θ, denoted by θ_a and θ_b, and two values of ϕ, ϕ_a and ϕ_b, and consider doing experiments with all four pairs of angles.

Now it is an experimental fact that whenever we measure the spins of both particles 1 and 2, in the *same* direction, we always find that they are opposite; if 1 is + then 2 is − and *vice versa*. As noted above this is what we expect from the fact that we have a state of total spin zero. Since we do not *know* whether we are going to measure the spin of particle 1, or along which direction if we do, it follows from the locality assumption that particle 2 must carry with it the value (+ or −) of its spin in any direction. Hence, with regard to the spin direction, there can in fact be no random element (this result is not required in the proof of the inequalities but it makes it easier).

We suppose that a given event, i.e. a given pair of particles, is completely characterised by a quantity we call H. It follows from the above paragraph that, among other things, H tells us the spins of the particles in any direction. For a particular H, suppose that the spin of particle 1 along a direction which we call θ_a is $S(H, \theta_a)$, and of particle 2 along a direction ϕ_a is $S(H, \phi_a)$. In each case we know that the S are either + or −. In table 10.1 we show the possible values of S for two directions of the spin of particle 1, θ_a and θ_b, and similarly two directions for particle 2, ϕ_a and ϕ_b. Every value of H is associated with one column of this table.

Now suppose that we could do the following four experiments successively, and with a particular value of H. In the first we measure the spins along the directions θ_a for particle 1 and ϕ_a for particle 2. We record only the product, i.e. $+1$ if the spins are equal and -1 if they are different. Call this product $P(a,a)$. In the other three experiments we do the same thing but with directions (θ_b, ϕ_b), (θ_a, ϕ_b) and (θ_b, ϕ_a), thereby recording numbers $P(b,b)$, $P(a,b)$ and $P(b,a)$. We now define the quantity

$$F(H, \theta_a, \theta_b, \phi_a, \phi_b) = P(a,a) + P(b,b) + P(a,b) - P(b,a). \quad 10.4$$

Note the signs; we add the first three and subtract the last. The P's will in general depend on the H's, and hence so will the value of F, a fact which we have indicated explicitly by writing H as one of the arguments of F. In table 10.1 we have calculated all the values of F, and we see that it is always either $+2$ or -2.

Table 10.1. Possible results of spin measurements.

θ_a	+	+	+	+	+	+	+	+	+	+	+	+	+	+	+	+
θ_b	+	+	+	+	−	−	−	−	+	+	+	+	−	−	−	−
ϕ_a	+	+	−	−	+	+	−	−	+	+	−	−	+	+	−	−
ϕ_b	+	−	+	−	+	−	+	−	+	−	+	−	+	−	+	−
F	+2	+2	−2	−2	+2	−2	+2	−2	−2	−2	+2	+2	−2	+2	−2	+2

Unfortunately we cannot do what we imagined in the last paragraph because we do not know H, and have no way of knowing that we are doing the four experiments with the same H. Instead, however, what we can do is to repeat the measurements, with each pair of angle settings, a large and equal number of times. In this way we average over all the columns in table 10.1, and obtain a quantity $\overline{F}(\theta_a, \theta_b, \phi_a, \phi_b)$ which is the average of $F(H, \theta_a, \theta_b, \phi_a, \phi_b)$. Of course we do not know how many times the positive and negative values will occur, since we do not know the distribution of the H, so we cannot calculate the average. However, an average of a set of numbers, all of which are either $+2$ or -2, is always between these two numbers, i.e. its magnitude is certainly less than or equal to 2. In other words we find, for the experimentally measurable quantity $\overline{F}$, the result:

$$-2 \leq \overline{F} \leq +2. \quad 10.5$$

This is one form of the Bell Inequality.

Although the input to the above result is apparently very small, the inequality so obtained is violated by the predictions of quantum theory. In fact, for particular values of the angles between the directions of measurement (denoted by θ and ϕ in the above discussion), the prediction for the quantity F defined above becomes $2\sqrt{2}$, which does not satisfy the inequality 10.5. Thus we have Bell's theorem: *quantum theory violates locality*.

Naturally this result led to a great deal of activity, which is still continuing. On the one hand, there have been many attempts to specify, much more carefully than in our treatment, the precise form of the conditions that are necessary in order to prove the inequalities. For an up to date review we refer to Redhead (1988). On the other hand, there were several experiments to test whether the real world agreed with the predictions of quantum theory, or whether it would violate them in such a way as to be consistent with locality. The best of these, done by Aspect's group in Paris, gave a very clear verdict: the predictions of quantum theory are correct. Further details of the experiments, and references to the original papers are given, for example, in Squires (1986).

Bell's theorem, and the subsequent experimental tests, are undoubtedly among the most exciting things that have ever happened in physics. The implications for other parts of physics, and even for things outside physics, may well be very profound and influential. One day, we might even understand them!

10.7   Time and quantum mechanics

We close this chapter with three rather vague and speculative comments regarding time. They are not necessarily connected, although all could have something to do with the locality issue of the previous section.

First, we note that in addition to non-locality in space we should really introduce something similar in time. We referred previously to wavefunctions of the form $\psi(\mathbf{x}_1, \mathbf{x}_2, t)$, which had two positions and one time. A quantity like this might be reasonable in non-relativistic physics, but not in a theory consistent with special relativity. Unfortunately it would take us too long to delve into a proper treatment of relativistic quantum theory (even to the extent that this exists), but it seems clear that we would then be more likely to deal with functions containing two times as well as two positions. This makes problems if we wish to speak of a system at some fixed initial time.

Next, we refer to the fact that discussions of the non-locality arising in EPR-type situations, as in the previous section, rely on the idea of "signals" being sent from one event to a later event. Implicitly we are here thinking in a dualistic way, i.e. we are assuming, first, the existence of a *free agent* which can perform some action at a particular place (A) and time T_1, and, secondly, that the effects of this action only reach another point (B) at a finite time later, i.e. at T_2 with

$$T_2 \geq T_1 + \frac{L}{c}, \qquad\qquad 10.6$$

where L is the separation between A and B and c is the velocity of light. (Here we have used the result of special relativity that signals cannot travel faster than light.) Now, in a totally deterministic world, there are, strictly speaking, no such free agents. The supposedly free actions in fact happen because of some earlier causes, and these causes could themselves communicate the required message to B, so that the effects at B could happen earlier than would be allowed by the above equation. Somewhat paradoxically, the idea of causality ceases to be significant in a totally deterministic world. Such a world just *is*, and, although various events are correlated, there is no obvious sense in which one part "causes" another. Examples of this sort of thing are very familiar of course. It may take many hours for the Sunday newspaper to reach my home, but I will already know the result it will give for a particular football match if I happened to be present myself at the game. (On the other hand we believe, surely correctly, that it is the results of the games that "cause" these results to appear in the newspapers.)

More relevant to us here is that the EPR apparatus, used to test the Bell inequality, is designed and set up long before the actual experiment is performed. The settings are not *really* chosen in a random way, so there is no rigorous demonstration of non-local effects. It has been suggested that there might be a way around this objection by using signals from distant galaxies, that have never been in causal contact, to fix the settings of the spin measuring devices. If such an experiment were possible, would it give disagreement with quantum theory? An alternative idea, being studied by a research student here in Durham, L Hardy, is that there might exist genuine free agents which are outside the physically determined world. Such free agents could be responsible for "mind-acts" affecting the settings in the EPR experiment. Assuming these are constrained by the Bell inequality, they would give rise to violations of quantum theory. (Experiments along these lines would be precise tests of a well defined type of dualism. Unfortunately, the time scales involved suggest that they would be very difficult to perform.)

The final remark is that the whole idea of a wavefunction being something that exists in space–time (i.e. being a function of x and t) may be false. The truth could rather be that there is no time dependence of the wavefunction, i.e. that at the level of the complete *wavefunction of the universe* there is no such thing as time! Such an apparently absurd suggestion is made, for example, in the work of Hawking (1984, 1988).

How is it possible that a wavefunction with no time dependence can describe the changing world of experience? The answer (if there *is* an answer) lies in the fact that a wavefunction contains many different physical situations, e.g. particle positions, etc, and in general these are correlated. Thus, for example, we might imagine that in those parts of the wavefunction where the radius of the universe is small (i.e. where it is peaked around small values), there is very little likelihood of finding particles close together in galaxies. On the other hand, in the parts of the wavefunction where the average value of the radius is much bigger, there will be a greater tendency for there to be galaxies; where it is bigger still we even find a reasonable likelihood of finding people, etc. Thus we might try to regard the radius of the universe as being like the hand on a clock. If we do, then the world changes with "time", i.e. with the position of the hand of the clock.

For those who can follow equations, it is amusing to see the simplest possible example of this (Englert, 1989). We consider a "world" of two free particles, with masses M and m, moving in one space dimension. We suppose that these particles are in an energy eigenstate, which means that the wavefunction satisfies the equation

$$\left(-\frac{\hbar^2}{2M}\frac{\partial^2}{\partial x^2} - \frac{\hbar^2}{2m}\frac{\partial^2}{\partial y^2}\right)\Psi(x,y) = E\Psi(x,y), \qquad 10.7$$

where the particle positions are denoted by x and y respectively. We also impose a "boundary condition" of the form

$$\Psi(x, y = 0) = e^{iKx}e^{-\alpha x^2}. \qquad 10.8$$

It is this boundary condition that introduces the correlation between the two particles.

We now want to regard the position of the particle at x as a sort of clock. Since it is a free particle we expect it to move at a constant rate, which of course will be related to the momentum ($\hbar K$) by

$$\frac{dx}{dt} = \frac{\hbar K}{M}. \qquad 10.9$$

This equation has introduced a "time" variable. In terms of this we can write equation 10.7 in the form

$$i\hbar\frac{\partial Y}{\partial t} = -\frac{\hbar^2}{2m}\frac{\partial^2 Y}{\partial y^2} + \left(\frac{K^2}{2M} - E\right)Y + \frac{1}{2M}\frac{\partial^2 Y}{\partial x^2}, \qquad 10.10$$

where we have defined $Y = e^{-iKx}\Psi$ as the wavefunction of the particle with mass m. Apart from the last term, which is small for large M, this equation is the time-dependent Schrödinger equation. Thus we appear to have produced a "time" from an equation which was independent of time.

I am not at all sure what this means. We have to wonder, for example, whether the quantity we have introduced has anything to do with time as it is experienced, and how it is possible to reconcile this model with relativity. On the other hand, it might be considered satisfactory that these sort of ideas seem to make time different from space. Even here however it is not clear whether this is so. If we can eliminate time as a fundamental quantity in this way, why can we not also eliminate one (or all) of the dimensions of space?

Chapter 11

What does
quantum theory mean?

11.1   The interpretation problem

As we have already stated several times, quantum theory is an
extremely successful theory; it agrees with observation even in situations where its predictions seem to violate what common sense
would suggest, and it allows us to calculate correctly a wide variety
of atomic properties, e.g. energy levels and scattering processes.
Although the calculations rapidly become extremely difficult in
practice, for all but the simplest cases, there are no ambiguities, at
least until we are required to include the effects of special relativity and of gravity. When, however, we ask how the theory *explains*
what is happening, or enquire into what it says about the external
world, the story is very different. Here we meet only confusion and
controversy! This is the interpretation problem of quantum theory,
a problem which has been the subject of many books and hundreds
of articles, offering a bewildering variety of "solutions". Although
the problem is as old as this century, it is not solved, and there
has recently been a resurgence of interest in it. We cannot hope
to give a complete account of all this work in the present chapter, but we will try to review the essential points and to describe
what appear to be the most plausible resolutions of the problem.
We can conveniently classify these into three groups (though other
classifications are possible: for example, Bell, 1987a, describes six
possible quantum worlds and Sudbery, 1986, offers nine different
interpretations).

The first group is characterised by the fact that quantum theory is *not (quite) correct.* The true theory contains small corrections to quantum theory, which make no measurable difference to the predictions for simple systems, but which, for more complex systems (such as all measuring apparatus), completely alter the results. Alternatively, it may be that there are additional effects, not included within quantum theory, which only play a role for certain types of system. It is not unreasonable to regard the "orthodox" interpretation of the theory as belonging within this class, because, although traditionally its followers have tended not to care about, or even to admit the need for, the extra corrections, the wavefunction describing the system does not always obey the Schrödinger equation, but has to undergo collapse. We discuss the orthodox interpretation more fully in section 11.2. Then, in section 11.3, we shall mention some specific suggestions that have been made for the correction terms.

In the next group of interpretations, quantum theory is considered to be *correct, but incomplete.* The wavefunction always changes with time according to the Schrödinger equation, but the wavefunction is not the complete description of the system. On the contrary, a complete description of a system at a given time requires, in addition to the wavefunction, a set of so-called *hidden variables.* This last expression is not really well chosen since, in the particular model we shall consider, and which we shall see is certainly the most natural, the hidden variables are in fact the familiar positions of particles. Indeed this model reintroduces genuine, classical, particles. The new quantum feature is that these particles are now subject to curious extra forces. All this is treated in more detail in section 11.4.

The final class of models go under the name of the *many-worlds* interpretation, although again, for reasons we shall discuss, I think this name is inadequate and even misleading as a description of the model. Here we meet perhaps the most bizarre description of reality that has been suggested on the basis of quantum theory. Nevertheless, this interpretation is the one which takes the actual formalism of orthodox quantum theory most seriously. It regards the theory as *correct and as complete as possible.* We discuss this more fully in section 11.5.

It is one of the fascinating features of quantum theory that, in choosing between a variety of possible interpretations, we are not faced with a selection of plausible alternatives. Rather it is the case that *all* interpretations, when we think about what they imply, are implausible. We find it hard to believe that the world really is like this. In the language used in section 10.1, all interpretations are "silly". Presumably, however, one, or perhaps one we have not thought of, which almost certainly would be even stranger than those we have, has to be correct. Here it is important to remind ourselves that, although we work within the framework of quantum theory, the problem is not merely a problem with a particular theory, it is a problem with the experimental observations of our world. Thus, for example, even though we might try to remove the difficulties from quantum theory, by asserting that the theory only applies to many identical copies of a system, i.e. to so-called *ensembles*, this does not help in solving our real problem: how do we explain observations?

11.2 The orthodox interpretation

Here we shall attempt to describe what is usually regarded as the orthodox interpretation of quantum theory. It will be realised, however, that such a term is by no means well defined. Most practising quantum physicists probably regard themselves as supporting this interpretation, but there would be a wide variety of statements about what it actually means! This of course is not surprising because the orthodox interpretation was originally developed, in the exciting early days of quantum theory, very much alongside of the formal, mathematical, description of the new theory and its applications. It originated in the writings of Born, Heisenberg and, supremely, Bohr in Copenhagen; from whence we have the alternative name: the "Copenhagen interpretation".

We shall describe the orthodox interpretation by listing some of its key features.

(a) An important aspect of orthodoxy in quantum theory is an implicit doctrine something like: *the theory works, don't worry too much about why, just go ahead and use it!* There is no doubt that such advice has proved wise and productive. It allowed, and continues to allow, many people to develop the successful applications of quantum theory to which we have often referred. It is not hard to see why, in the early days, the rapid, exciting, progress being made by the theory, and its ability to predict hitherto mysterious effects, discouraged too much concern with the foundations. To pose too many questions would be to seem too conservative, too lacking in the ability to learn and accept new things. There was the feeling that the interpretation questions belonged to some hazy area of "metaphysics". In addition, those who did wonder, were deterred from spending too much time on the problem by a vague feeling that it had been solved by Bohr and his collaborators. In the words of one of the leading theoretical physicists of the second half of this century, Murray Gell-Mann: *The fact that an adequate philosophical presentation has been so long delayed is no doubt caused by the fact that Niels Bohr brainwashed a whole generation of theorists into thinking that the job was done fifty years ago* (Gell-Mann, 1979).

(b) The general lack of interest in the nature of the reality which lay behind the formalism of quantum theory was supported by the prevailing philosophical doctrine of the period, which was positivistic and anti-realistic. Similarly, the fact that quantum theory could not readily be seen in realistic terms encouraged these trends in philosophy. Even in 1958 Bohr could write: *there is no quantum world, there is only an abstract quantum physical description* (Bohr, 1958, p.6). Quantum theory gave the correct answers; what right had we to expect anything more from a theory?

(c) Bohr made much use of the word, and concept of, *complementarity*. Under some circumstances it is possible to describe a "particle" using obviously particle-like variables, e.g. position and velocity. On the other hand there are times when it is more appropriate to use wave-like properties to talk about the particle. These two concepts are perfectly acceptable of course, and we are familiar with this sort of thing in ordinary life: whether I describe a book in terms of its content or its thickness depends upon whether I want to read it or simply use it to fill a slot on a shelf. The complementarity idea in quantum theory, however, has to accept that the two descriptions are incompatible; although both are valid in some respects, they cannot be fitted together, *and we have no right to expect that they should.* There is no comprehensive picture of the system that contains both complementary descriptions. At this stage the orthodox interpretation begins to sound like "mumbo-jumbo", a collection of empty words that indicate only that we really do not understand quantum theory. The usual idea behind two descriptions being complementary is that they should not be contradictory!

(d) Although the beautifully elegant formalism of quantum theory, presented in all its splendour by Dirac (1930), required the addition of an extra, clumsy, rule, i.e. that something rather vague happens when a "measurement" is made (recall section 10.5), it was possible in the early days to pretend that measuring apparatus could be described by classical physics. Perhaps this was rarely made explicit, but it was only with a great sense of pioneering that physicists were applying quantum theory to *microscopic* objects, so they were not in any way embarrassed by simply using classical ideas to describe measurements (which obviously always involve large systems). Such classical systems "of course" always had fixed values of position, etc; *they* were not subject to the wave ambiguity of microscopic objects.

To some extent, all this remained true even after von Neumann (1932) had shown that *quantum systems could never make measurements of the type required by the above rule.* (We discuss this further in the next section.) It was still possible to escape with some vague feeling that big objects did not obey quantum theory. From a late-20th-century viewpoint, such an escape no longer seems to be available. It might be true, but we want to know why, and to know what laws they *do* obey.

(e) In so far as it is allowed to be a sensible question, there is uncertainty in the orthodox interpretation about the status and meaning of the wavefunction. On the one hand, the evidence of interference implies that the wavefunction is a part of external reality, e.g. like a water wave, otherwise we do not have any answer to the question of what is doing the interfering. It is worth emphasising here that, in all the interpretations we shall discuss, this fact, the basic "ontological reality" of the wavefunction, remains true. On the other hand, there is the unpleasant feature of this wavefunction that it seems to undergo sudden jumps when "measurements" are made.

We must remind ourselves why these jumps ("collapse" or "reduction" of the wavefunction) seem to be required. Suppose, for example, that in the barrier experiment of section 10.3, we *observe* that the particle has passed through the barrier. What this might mean in practice is that we see a flash on a suitably placed fluorescent screen. Now, in some sense at least, this must mean *more* than the fact that I have *observed* the particle to have gone through; it must mean that it *has* gone through (we shall be less confident of this in section 11.5). Since the wavefunction is, within orthodoxy, the complete description of the physical situation, the wavefunction itself must therefore show that the particle has passed through. It must somehow have reduced from the form shown in figure 10.7(b) to that of figure 11.1. Further evidence for this comes from the fact that I can make a later measurement, or "you" can measure whether the particle has passed through, and the answer must always be that it has. That is, as soon as I have made my observation that the particle has passed through, the probability of its being on the other side has become zero. According to the standard assumption of quantum theory this means that the wavefunction on the other side must also be zero, in accordance with the figure.

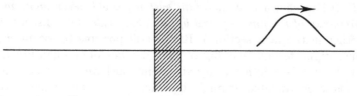

Figure 11.1. Here we see a possible form of the collapsed wavefunction, obtained from the wavefunction of figure 10.7(b), when the particle has been observed to be transmitted.

This type of effect is of course very easy to understand in a classical probability situation, where the wavefunction does not represent anything physical, but is merely a statement about *our knowledge* of the physics. We might, for example, be told that an object has a 50% chance of being in one box and a 50% chance of being in another. This would be described by a probability distribution function having two equal peaks, one at each box. If we look into one box and find the object present, then the probability function immediately becomes zero in the other box. All that has changed however is our knowledge; nothing has happened to the real world since the object already was in one box, even before we looked.

It is of vital importance to realise that the quantum theory situation is very different to this. The wavefunction *is* the system.

It is not true to say that before we looked the particle really had passed through, or not. If it had then there could be no interference (see also section 11.3).

It is interesting to remark that Schrödinger, who invented the wavefunction, was very unhappy with its being expected to collapse. Elsewhere (Squires, 1986) I have written: "We recall that it was he who introduced the equation that bears his name, and which is *the* practical expression of quantum theory, with solutions that contain a large proportion of all science. In 1926, while on a visit to Copenhagen for discussions with Bohr and Heisenberg, he remarked: *If all this damned quantum jumping were really to stay, I should be sorry that I ever got involved with quantum theory*".

This, then, is the orthodox position. It is incomplete. In particular, it does not tell us what sort of things can cause the effects of "measurement", or how, or when, etc. We discuss possible ways to answer these questions in the next section. If we are happy to ignore them we are accepting what Bell (1987a) calls the "pragmatic" interpretation, the "romantic" version of which tries to make a virtue of the inconsistencies (see (c) above). Readers who require a fuller account of the Copenhagen interpretation, and further references, should consult Stapp (1972).

11.3 What can make a quantum measurement?

In this section we shall explain why quantum systems, i.e. systems described by the Schrödinger equation, cannot themselves make measurements in the sense that we have discussed above. We shall then introduce a few of the suggestions that have been made to change the laws of physics so that measurement becomes possible. The discussion of this section is, necessarily, somewhat more mathematical than most of the book. However, the equations are merely symbolic ways of stating what is explained (I hope adequately) in the text.

We shall consider a system which has effectively two degrees of freedom. To be definite we take an electron, for which the spin in a given direction, which we choose to be vertical, can take one of two values. We denote these by $+$ and $-$, respectively. The most general possible wavefunction describing the system then contains a certain proportion, say α, of the $+$ state and a certain proportion,

β, of the $-$ state. We introduce the abstract notation of Dirac, in which wavefunctions are written in the form $|\dots>$, so we can express such a wavefunction as

$$|\Psi> = \alpha|+> + \beta|->. \qquad\qquad 11.1$$

If we make a measurement of the spin in the chosen direction we will obtain $+$ or $-$ with probabilities $|\alpha|^2$ and $|\beta|^2$, respectively. This statement is essentially the content of equation 11.1. Note that, since we will obtain *some* result when we make the measurement, the two probabilities have to add to unity, i.e. $|\alpha|^2 + |\beta|^2 = 1$.

Suppose that the spin is to be measured with a Stern–Gerlach device. This consists of a varying magnetic field, in the vertical direction, which causes the electron to be deflected, say, upward if the spin is $+$ and downward if it is $-$. The precise details of the apparatus, or why this happens, need not concern us here. To describe the electron after it has passed through this apparatus we must introduce into the wavefunction a variable associated with the direction in which the electron moves. If the electron had a definite state of spin, say $+$, then the complete state after the measurement would be $|+, up>$. Similarly, if it was $-$, then the final state would be $|-, down>$. All this is simply a statement of what we mean by a good measuring apparatus. The final wavefunction in the case of an electron in the state described by equation 11.1 is now a direct consequence of the linearity of the Schrödinger equation. It will have a piece in which the electron has $+$ spin and is moving upwards, together with a piece in which it has $-$ spin and is moving downwards. In symbols:

$$|\Psi> = \alpha|+, up> + \beta|-, down>. \qquad\qquad 11.2$$

The first thing to note about this is that we have in one sense made a "perfect" measurement, the sort of measurement only a *theoretical* physicist (who has not done a real experiment since he was an undergradute) can make! That is, the correlation between the spin direction and the direction of motion is complete; there are no terms containing $|+, down>$, for example. However, in a more important sense, namely, the sense in which the term is used in quantum theory, we have not really made a measurement at all. The above expression still contains the electron with both directions of spin. A proper measurement, on the other hand, should have resulted in one particular answer being found, in which case the wavefunction should only contain one term, i.e. it should be **either**

$$|\Psi >= |+, up >, \qquad 11.3a$$

or

$$|\Psi >= |-, down > . \qquad 11.3b$$

Of course we could say that the reason why we have not made the measurement is that we have not "observed" whether the electron went up or down. To discuss this, we need to introduce a detector of some sort which reads *on* if the particle goes up and *off* if it goes down. (The nature of the detector is of course irrelevant to the discussion.) The wavefunction must now contain a factor describing this detection apparatus, and initially it will describe this to be in, say, the *off* state. It is then easy to see that after the electron has passed through the detector, and provided this is doing its job properly, we have as the new wavefunction, which readers should now be able to interpret,

$$|\Psi >= \alpha|+, up, on > + \beta|-, down, off > . \qquad 11.4$$

The same remarks apply again; we have not succeeded in making a proper quantum measurement with a unique result. Indeed it is clear that in this way we are never going to obtain a wavefunction that does not contain both possibilities. We can easily understand why this is so. We are using a deterministic equation to tell us the wavefunction, so we will never be able to obtain a result which is one thing **or** another. We have demonstrated the very important result that systems described by quantum theory cannot make measurements.

A wavefunction that contains both (more generally several) possibilities in a linear superposition, like that in equation 11.4, is called a *pure* state. On the other hand, a wavefunction that contains one *or* the other, like that in equation 11.3, describes a system in a *mixed* state. The result that we have shown can be stated by saying that a pure state cannot, under the time evolution given by the Schrödinger equation, become a mixed state. We emphasise again that the two states describe genuinely different physics, because, in principle at least, there can be interference between the parts of a pure state. In *practice* there are difficulties in actually observing this interference except for very simple, microscopic, systems. Indeed, many authors have shown how the effect of the environment, which we have ignored in the above, ensures "decoherence" and so prevents interference for all practical purposes. It does not, however, solve the measurement problem.

It follows from this discussion that if we wish to reduce the wavefunction then we must somehow change the theory. The changes that we can make must satisfy several conditions:

(1) They must cause the wavefunction to reduce in a way that is compatible with the probabilistic measurement rule of quantum theory. Thus, in the above example, we must obtain $|+, up, on >$ with probability $|\alpha|^2$, and $|-, down, off >$ with probability $|\beta|^2$.

(2) The reduction must be compatible with what is being measured, i.e. if we measure a particular quantity, say a position, then the reduced wavefunction must be such as to have a definite value of that quantity. (Mathematically all measurable quantities correspond to operators in quantum theory, and what this means is that the reduced wavefunction must be an eigenvector of the appropriate operator.)

The sort of problem that is raised by this requirement can be seen if we think again of the spin example above. Instead of expressing the state in terms of the spin values used in equation 11.1, we can choose some other direction to define the spin states. If we denote the new states by $|+ >'$ and $|- >'$ then the general state in equation 11.1 can be written as

$$|\Psi >= \alpha'|+ >' + \beta'|- >' . \qquad 11.5$$

Note that equations 11.1 and 11.5 describe the same state; the different descriptions arise from the use of different bases (just as we can use different coordinate systems to express an ordinary vector in space). Clearly we cannot simply say that an electron wavefunction reduces to one of definite spin projection, because we would not know whether it would have to reduce to one of the states on the right-hand side of equation 11.1 or of equation 11.5.

(3) The rate of reduction must be carefully balanced. It must be sufficiently quick to accord with the fact that when we make an observation we normally obtain an apparently instantaneous result. On the other hand, it must be sufficiently slow, i.e. the

correction terms in the Schrödinger equation must be sufficiently small, that the good predictions of quantum theory, e.g. for the energy levels of atoms and the interference of neutrons, are not destroyed.

There is one important feature of all measuring systems which gives us some hope of being able to satisfy these conditions, namely, they are *very large*, i.e. they are *macroscopic* objects containing a large number of atoms. Here we recall that 1 kg of matter contains about 10^{25} atoms. Thus, provided the effect is approximately proportional to the number of atoms, or to the mass of the object, then we can include terms which have a negligible effect on microscopic systems, with a few atoms, but which are dominant for macroscopic apparatus. Hence it is not hard to arrange things so that the collapse takes longer than the age of the universe for a single particle, but is essentially instantaneous for any practical measuring apparatus.

Condition (2) can also be satisfied because it is reasonable to assume that all measurements are ultimately measurements of position of some macroscopic object. Thus, we require that the wavefunctions of macroscopic objects should reduce rapidly to wavefunctions corresponding to definite positions (with of course some small "error"). For example, in the above measurement of the spin of an electron, the states "on" and "off" might correspond to different positions of a pointer. Then, when this reduces to a *particular* position, the whole wavefunction must become either $|+, up, on >$ or $|-, down, off >$. The electron spin becomes definite only in the direction corresponding to the setting of the measuring apparatus, in particular, the direction of the Stern–Gerlach magnetic field.

Clearly there must be some sort of "stochastic", or random, term in the modified Schrödinger equation, otherwise the result would inevitably be deterministic, which is not what is required. Also, in order to obtain agreement with condition (1) above, the new equation must be non-linear.

As a simple example of a mechanism, or perhaps rather a prescription, which has the necessary properties, we describe the interesting suggestion of Ghirardi, Rimini and Webber (1986, GRW), explained in a simple form by Bell (1987a, p.201, 1987b). The idea here is that in any time interval there is a certain probability that a particle will "jump" to a definite position, with the relative probability of particular positions chosen according to the magnitude-squared of the wavefunction at that point. These probabilities are so small that, for a single particle, a jump will effectively never occur (hence we get perfect interference patterns with neutrons, etc). However, in a macroscopic object, the time taken for one jump to happen is reduced by a factor equal to the inverse of the number of particles. Thus, provided it is arranged that when one particle jumps it reduces the whole wavefunction (see below), then such reduction is effectively instantaneous. For interested readers we shall now give a little of the relevant mathematics. Others should skip the next three paragraphs.

Consider a system of N particles, described by a wavefunction $\Psi(\mathbf{r}_1, \mathbf{r}_2, \ldots, \mathbf{r}_N)$. In every unit of time there is a probability N/τ, where τ is a suitable chosen free parameter, that the wavefunction has jumped to a new wavefunction Ψ', given by

$$\Psi' = \rho(|\mathbf{R} - \mathbf{r}_n|)\Psi(\mathbf{r}_1, \ldots)/\phi_n(\mathbf{R}), \qquad 11.6$$

where n is randomly chosen from the set 1 to N. The function ρ is peaked about the argument zero, with a width that is another free parameter of the theory. Because this width is non-zero, the reduction occurs to *approximate*, rather than exact, eigenstates of position. The normalisation factor ϕ is defined by

$$|\phi_n(\mathbf{R})|^2 = \int d^3\mathbf{r}_1 \ldots d^3\mathbf{r}_N|\rho(|\mathbf{R} - \mathbf{r}_n|)\Psi|^2. \qquad 11.7$$

This function also determines the distribution of the centres of collapse, which are chosen at random, with distribution given by $|\phi_n(\mathbf{R})|^2$.

For the time constant τ, a value around 10^{15} s or 10^8 years is suggested. Then it is clear that no effect will be seen in a system consisting of a few atoms. On the other hand, for a system with 10^{22} particles, at least one will have jumped in a time of about 10^{-7} s. The width of the peak in ϕ is taken to be about 10^{-5} cm, which is small enough to describe the localisation of macroscopic objects.

In order to see how the jumping of one particle causes the whole wavefunction to reduce, we might consider again the case of measurement of a particle spin. Suppose that the *on, off* states of the recording device correspond to two different positions of a pointer. Before reduction the wavefunction has the form given in equation

11.4. However, in a time of the order of a microsecond or less, one of the atoms of the pointer will have jumped, i.e. the wavefunction will be multiplied by a factor, $\rho(|\mathbf{r}_1 - \mathbf{R}|)$. Here we have chosen particle 1 to be the particle that jumped. Thus, instead of equation 11.4, we will have

$$|\Psi> = \alpha\rho(|\mathbf{r}_1 - \mathbf{R}|)|+, up, on > +\beta\rho(|\mathbf{r}_1 - \mathbf{R}|)|-, down, off >.$$
$$11.8$$

Now, $\phi_n(\mathbf{R})$, from equation 11.7, is small except in the neighbourhood of the *on* or *off* pointer positions. Suppose, for example, that the random process selected a value of $\mathbf{R}$ lying in the *on* pointer position. Then, provided the width of the peak in ρ is small compared with the separation of the pointer positions, there will be no "overlap" between ρ and the other pointer position, and hence there will be no values of the coordinates of the particle that has jumped, i.e. of $\mathbf{r}_1$, for which the second term in equation 11.8 differs from zero. If $\mathbf{r}_1$ is close to $\mathbf{R}$ then the wavefunction $|off >$ is zero; on the other hand, if it is near the *off* pointer position, then ρ will be zero. Thus the second term in equation 11.8 is always zero and can be dropped. We have accomplished the required reduction of the wavefunction, which now shows only one spin direction.

Further details of this idea can be found in the original articles noted above. The method seems to work and not to give rise to any problems elsewhere. Nevertheless it is somehow hard to believe that nature really does behave in this strange fashion, so perhaps the idea is best seen as a means of illustrating the contortions that are required if we are to devise a method of reducing wavefunctions. It is unpleasant to have to introduce two new fundamental scales, the time scale, τ, and the width of the function ρ into nature, although it is hard to see how any method of producing wavefunction reduction can avoid this.

Of course the GRW prescription that we have just described does not really provide us with a dynamical theory of wavefunction reduction. For that we would expect not to have to *assert* that jumps happen, we would rather expect to be able to see how and why they happen, in other words we would like the behaviour to come from some sort of equation, presumably a modified form of the Schrödinger equation.

Pearle (1976) showed how it is possible to add extra, non-linear, stochastic, terms to the Schrödinger equation, so that the wavefunction would always reduce to a form that corresponds to a definite value of one particular measurable quantity. There was no natural way of choosing which quantity, and this model did not have the advantage of giving more rapid reduction for macroscopic systems. A recent attempt to combine some of the features of the GRW and the Pearle model is given in Pearle (1988) and in Ghirardi *et al.* (1989b). There have also been several recent attempts to suggest physical origins for the extra, stochastic, terms, e.g. Diosi (1989), Ghirardi *et al.* (1989a), Gisin (1989), Kent (1989), Percival (1989) and Stapp (1989a). There are many other, less well developed, attempts to specify the important features of a "measurement" that allow it to reduce wavefunctions, e.g. Bussey (1984, 1986) and Maxwell (1988) (see also Squires, 1989a).

Models for wavefunction reduction are a lively topic at the present time. It will be of great interest to see how they meet the challenge of special relativity on the one hand, and the requirements of EPR-style non-locality on the other. (See Bell, 1987a, p.201, Pearle, 1989 and Squires, 1989b, for some tentative remarks on these questions.) Another point which needs further consideration in this type of model is that of time-reversal invariance. The jumps in the GRW model clearly violate this invariance, so maybe they could provide the time direction manifest in our experience. On the other hand, the stochastic equations which purport to give similar effects seem, at first sight, to be invariant under reversal of time. See Chung and Walsh (1969) and Williams (1979) for discussion of time-reversal and stochastic processes.

11.4   Consciousness and wavefunction reduction

At the beginning of the previous section we showed that systems described by quantum theory change with time in a deterministic way, and that there is no sign of the probabilistic reduction of the wavefunction associated with obtaining a unique result of a measurement. The reason why we are not satisfied with this is that we are able to make measurements that do in fact yield a unique result. We know this because, provided the measurement is suitable, *we are aware of only one result.* If it were not for this awareness then the quantum theory description is perfectly satisfactory. We know that wavefunctions, as in equation 11.1, containing a super-position of both spin values (or particle having passed through and particle having been reflected), are correct (because interference is possible). The linearity of quantum theory then requires us to ac-cept that equations 11.2 and 11.4 are also valid. The complete system, including the apparatus, is described by a superposition of states. Let us now go further and put *ourselves* into the physical process. In particular I will put "Me" into the wavefunction. Only two states of my brain are relevant, namely, Me^+, and Me^-, cor-responding to my having observed the particle to have had + or − spin respectively. (It is of course not important how I actually make the observation, i.e. which piece of the measuring apparatus I choose to look at.) The complete wavefunction now has the form:

$$|\Psi> = \alpha|+, up, on, Me^+ > + \beta|-, down, off, Me^- > . \qquad 11.9$$

At this stage, and only at this stage, the wavefunction is apparently unacceptable, because it fails to describe my experience of one result.

A possible consequence of this sort of reasoning is that *wavefunc-tion reduction happens only when a conscious mind becomes aware of the result*. Among the eminent physicists who have considered this conclusion are von Neumann and Wigner (see, for example, Wigner, 1962). It does not immediately commit us to any particu-lar position on whether the relevant properties that allow conscious mind to reduce wavefunctions are *physical*, e.g. particular struc-tures or new types of particle that require different types of term in the Schrödinger equation, or whether they are more properly thought of as being beyond physics. Nevertheless it seems more naturally to lead to the latter hypothesis. We would then tend to say that *physics* does not have reduction of the wavefunction, i.e. physics is the Schrödinger equation.

It is necessary that we pause here to realise what all this im-plies. Suppose, for example, that we use a photographic plate to record whether the electron goes up or down in the above experi-ment to measure an electron spin. As long as the wavefunction is unreduced, it contains a piece where the plate is blackened in one place and a piece where it is blackened in another. We can put the plate away in a drawer and keep it for years, and it remains unsure where the track lies. Only when somebody *looks at it and becomes conscious of a track* does it actually become one thing or the other. Is this an acceptable description of reality?

Following Schrödinger's famous example, we can make an even more incredible story if we imagine that one electron path passes through an apparatus that is so designed as to fire a gun that kills a cat. Thus the wavefunction will contain a cat that is dead and a cat that is alive. Only when the state is observed by a conscious mind does it make sense to say that the cat is either dead or alive; until then it is in a superposition of being both dead and alive! (Of course, if the cat itself is conscious, then presumably it can decide for itself whether it is dead or alive. But would it want to take the risk of looking?)

In this scheme, the world as we know it is very much a product of conscious mind. Without conscious mind, for example before there were conscious beings on the earth, unreduced wavefunctions were the full extent of reality, no particle had "decayed", no particles had been scattered through a given angle, indeed there were not really any particles at all. Even such things as the "vacuum state" of the universe, which is crucial for determining whether the universe is a suitable place for life to develop, would not be fixed without having being observed. Thus it would be by observation that we would have created the conditions that are suitable for us to exist: observations would have created the conditions necessary for there to be any observations! All this is clearly very relevant to the Anthropic Principle, which we already mentioned in section 4.6, and which we shall meet again.

It is probably fair to say that most members of the physics community would reject the basic idea of this section. Their reasons would be based more on prejudice than sound argument, and the proportion of those who would reject it would be much smaller if we considered only those who had actually thought carefully about the problems of quantum theory (the cause–effect relation could, of course, go either way here). We shall introduce a variant of the idea in section 11.6, and discuss it further in the next chapter.

11.5   Hidden-variable models

Perhaps the best way of introducing the ideas of this section is to ask whether, in section 10.1, we gave up the idea of classical particles, following well defined trajectories, too easily. Why were we apparently forced to do this? The argument depended on the implicit assumption that a particle going along one path could not "know" whether or not the other path was open, and so it could not be affected by the other slit. Provided we limit ourselves to all known forces, this assumption is certainly valid. We could, however, take the view that the apparent interference phenomena prove that there are some new forces which act on the particles, and which *do* depend upon the situation of both paths, e.g. upon whether the slits are open or shut. The question is then whether we can find some rule for these forces, such that, when they are put into Newton's law of motion, they affect the particle trajectories in such a way as to give the appearance of interference. Of course, we must also remember the lack of determinism. Some particles of a given energy will pass through a barrier, and some will not, so if we are to describe their motion by Newton's laws they must experience different forces, even though the physical situation is the same. How can this be?

The key to the solution of these problems, given in the de Broglie–Bohm hidden variable theory, is to *add* to the classical description of the physical system the quantum theoretical wavefunction. This wavefunction does depend upon whether or not the two paths are open, and, because of its changing relation to the trajectory, it can give rise to the apparent lack of determinism. (It is here only "apparent" because if we knew the wavefunction and the position and velocity of a particle at a particular time we could exactly calculate the trajectory.)

It requires a little mathematics to see how we can make this model agree with quantum theory. First, we note that the predictions of quantum theory are *statistical*. That is, if $|\psi(x,t_0)|^2$ represents the probability of finding a particle, moving in one dimension, at position x at time t_0, then $|\psi(x,t_1)|^2$ is the probability of finding it at x at time t_1. In the de Broglie–Bohm model we want these functions to represent actual distributions of classical particles. The particles therefore have to move so that if $|\psi(x,t_0)|^2$ is their distribution at t_0, then $|\psi(x,t_1)|^2$ is their distribution at the later time t_1. It is a simple matter to obtain from this requirement a unique expression for the velocity of the particles, namely:

$$v(x) = Re \left\{ \frac{-i\hbar}{m} \frac{\partial \psi(x,t)}{\partial x} \bigg/ \psi(x,t) \right\},$$ 11.10

where m is the mass of the particles and Re means we take the real part of the following expression. If we recall the caption to figure 10.6, where we saw that the operator corresponding to the momentum $p(= vm)$ is given by $p = -i\hbar \frac{\partial}{\partial x}$, then we can write this equation in the form

$$mv(x) = Re \left\{ \frac{p\psi(x,t)}{\psi(x,t)} \right\},$$ 11.11

which makes it look very reasonable. (Note of course that in this equation p is an operator so we cannot simply cancel the factors of ψ.)

The acceleration of the particles can then be found by differentiating the velocity with respect to time, and using the Schrödinger equation to eliminate the derivative of ψ. This leads to the equation

$$m\frac{dv}{dt} = -\frac{\partial V}{\partial x} - \frac{\partial Q}{\partial x},$$ 11.12

where Q is defined by

$$Q = \frac{1}{2m} Re \left\{ \frac{p^2 \psi}{\psi} - m^2 v^2 \right\}.$$ 11.13

This is the usual form for Newton's law of motion, where V is the potential, except for the addition of the last term which is the extra, quantum, "force", dependent, as we see, upon the wavefunction.

This quantum force is the only difference between the quantum world, as described by the de Broglie–Bohm model, and the world of classical physics. The model is totally realistic and deterministic; the description it gives of the external world is very close to our (classical) perception of the world; it agrees with all the successful predictions of orthodox quantum theory; it does not suffer from any vagueness; and it requires no additional interpretation.

In spite of all these advantages the de Broglie–Bohm model is not highly regarded in the physics community. There are some sociological reasons for this. For a long time there was also a vague feeling that the possibility of a deterministic hidden variable theory had been ruled out by a theorem of von Neumann. In fact Bell showed that this theorem is irrelevant unless certain physically unreasonable conditions are imposed (see Bell, 1987a, for a careful discussion). The story of this theorem is an interesting example of the fact that theorems belong to mathematics, and that *there are no theorems in physics*. I made a remark similar to this at a recent talk in the University of Kent, and was then reminded that I had already utilised Bell's theorem (see section 10.6). Of course this theorem is another example of the same point: the world of physics does not satisfy its assumptions.

There are several genuine reasons why this particular hidden-variable theory is hard to accept. The extra quantum force is a very peculiar object. It gives agreement with experiment only if suitable boundary conditions are imposed, and then only at the expense of requiring that particles follow very strange trajectories (e.g. Philippidis *et al.*, 1979). Also, it is independent of the positions of the other particles in the system, unlike all the other known forces. Its relation to the wavefunction is such that it can be arbitrarily large even where the wavefunction is very small. If the system is in a bound energy eigenstate, then the quantum force cancels the other forces and the system becomes free. Of course none of these points are totally convincing; our normal expectations of what is reasonable may just be wrong.

One aspect of hidden-variable theories that is sometimes, mistakenly, assumed to forbid them, but which certainly causes them problems, is that, being so explicit, they reveal the non-locality of quantum theory in a stark and unpleasant way. We can see this from equation 11.13, if we imagine the wavefunction to refer to several particles. Then the force on one particle at a certain point will depend upon what is happening at many other points of space. It was this explicit non-locality that led Bell to remark that the model "completed" quantum theory, and therefore solved the EPR problem (section 10.5), in the way that Einstein would least have liked. It also provided the inspiration for Bell's theorem, which is sometimes stated in the form that no *local* deterministic theory can agree completely with quantum theory. In my opinion, however, it is not reasonable to use this as a reason for rejecting non-local models. The non-locality is a property of the world, and is not just brought in by the model.

Another reason why some are unhappy with hidden variables is the fact that the theory is "dualistic" in a strangely asymmetric way. It contains the familiar world of classical particles, moving according to Newton's law of motion. In addition it contains the world of quantum-theoretic wavefunctions. These exist and evolve in time totally independent of, and unaffected by, the existence and motion of the particles. The only interaction between the two worlds is due to the quantum potential as described above. Thus the quantum world affects the world of particles, but there is no effect in the opposite direction. (Note that the Schrödinger equation does not "know about" the positions of the particles.) Physics has taught us to mistrust such a state of affairs.

The situation is made even stranger when we realise that the two worlds have independent initial conditions. There is no reason why the wavefunction at the time of the big bang, say, should have any particular relation to the positions of particles. Even the average energy, as determined by the wavefunction, could be different to that obtained from the particle positions and velocities. Similarly, when we do an experiment we never really *know* the wavefunction, so there is no possibility of predicting its effect on the particles. In principle, at least, these effects could be arbitrarily large.

Of course in this model, as in all models where there is no re- duction of the wavefunction, there is an enormous "redundancy" in the physical world. Suppose, for example, that in a Schrödinger's cat experiment, which of course in the de Broglie–Bohm model is totally deterministic, the cat actually does not die. The part of the wavefunction in which it did die is still present and, in some sense, its future consequences will all be worked out in time. Thus the wavefunction will contain the possibility of the owner of the cat buying a new kitten, to replace the dead cat, etc. All these "other worlds" exist in the wavefunction. They are not real only if we restrict our concept of "reality" just to the hidden variables, i.e. to the particle positions.

There are considerable difficulties in trying to make the de Broglie–Bohm model compatible with special relativity, but I do not think these should necessarily be held against it because it is well known that a satisfactory combination of relativity with any form of quantum theory has proved very elusive. Attempts at rel- ativistic versions of hidden variable theories have been given by

Baumann (1986) and by Bell (1987a, p.173). Further discussion of some of the implications of the model, and references, are given in Bohm and Hiley (1984).

Finally there is one remark we should make which is particularly relevant to the topic of this book. As we have noted the model of this section is very close to classical physics. It says that the only modification that quantum theory, the physics of the 20th century, makes to classical physics is to add a new, very mysterious, non-local, "force". This force affects the way particles move, but is not itself caused by particles. From the point of view of 19th century physics, it comes from somewhere outside physics. Would any physicist of the 19th century have believed that such a thing could possibly exist?

11.6   The many-worlds interpretation of quantum theory

This interpretation began with an article of Everett (1957), although the name actually came later. Everett had been interested in the possibility of applying quantum theory to the universe, and using such ideas as the "wavefunction of the universe". (Since quantum theory was designed for, and has only been tested for, microscopic systems, this may well seem a dangerous extrapolation.) If we really do include everything in the wavefunction then there is nothing outside to make an observation, nothing that can reduce the wavefunction, etc. Hence the interpretation problem seems to become even more intractible. In response to this Everett tried the assumption that wavefunctions do not reduce, ever. Explicitly he asserted:

> *The wavefunction changes with time only and always in accordance with the Schrödinger equation.*

Apart from this statement, which as we saw in section 11.3 forbids wavefunction reduction, he followed orthodox quantum theory.

Clearly of all models this is the most economical in ideas and concepts. We implicitly dismissed it earlier because it apparently does not accord with the facts. Let us look again at the argument. It depended on equation 11.9, which we reproduce here:

$$|\Psi> = \alpha|+, up, on, Me^+ > + \beta|-, down, off, Me^- > .   11.14$$

In the discussion of section 11.4, below the equation, we claimed that "the wavefunction is apparently unacceptable, because it fails to describe my experience of one result". Is this true? If I am happy to leave an electron in a state which is a superposition of having gone through a barrier and not having gone through a barrier, or a photographic plate in a superposition of having tracks in different directions, or even a cat in a superposition of being both dead and alive, why cannot I allow my brain to be in a similar sort of superposition? The reason that I want to give comes back to my experience, which is that I see one result. But is this fact incompatible with equation 11.14? The answer to this begins to look as though it depends upon what I mean by experience (which means that it could be very relevant to our topic, and also that it will *not* be simple to obtain). If I imagine that my consciousness is able to survey the whole scene, that is to look at the wavefunction, then I would clearly be unhappy because I would see a wavefunction that contains nothing corresponding to my experience, i.e. nothing in any way distinguishing the result of which I am aware, from that of which I am not aware. It wants "me" to be in a situation of being unsure of the result, whereas I know I am not. In fact, however, I cannot do such a survey; I am not outside the wavefunction, rather I am part of it, and the Me^+ that has seen one result is totally unable to be aware of the Me^- that has seen the other. This last fact is a simple consequence of quantum theory. Actually it is not quite correct, because in principle the two Me's could interfere. As we have already noted however this is an utterly negligible effect for macroscopic systems (certainly for people). Basically, then, reality contains two Me's. The world (or my world?) has split into two.

This of course is the origin of the name "many worlds". The world appears to split into two (in general, many) branches, corresponding to the different results of observation. Of course, in fact, by writing the wavefunction in an expansion like that above, we do not imply that anything has happened to the world. The expansion is just a convenient way of writing the unique wavefunction, and there are many other expansions that we could have used. This is why I believe that the name is liable to be misleading, and why it has been suggested (e.g. Albert, 1986, Squires, 1987) that something like "many views of one world" might be better.

There are several attractive advantages that follow from the Everett assumption. We do not need to worry about what types of system can make observations (reduce wavefunctions). We do not have to put extra terms into the Schrödinger equation. It is possible to discuss quantum theory without having to introduce the idea of an "observer". Note also that, in spite of our not having had to reduce the wavefunction, we do not run into any problems when, for example, "you" also make a measurement on the same system. To see this we introduce you (denoted by Y) into the wavefunction, which then becomes

$$|\Psi> = \alpha|+, up, on, Me^+, Y^+ > + \beta|-, down, off, Me^-, Y^- > .$$

11.15

It is clear that the Me that has seen $+$ can only communicate with the You that has seen $+$, and *vice versa*. Hence "we", that is both we's, will always agree on the result.

The non-locality problem is resolved, or at least partially resolved, in the many-worlds model. The Schrödinger equation is a classical wave equation, albeit in many variables, and it is local. The non-locality associated with reduction of distant wavefunctions is not a problem, because it does not happen.

As we noted in section 11.5, all parts of the unreduced wavefunction exist and evolve for all times. However, in contrast to the case with the hidden-variable models, all parts now are equally real (in the ontological sense). *The world that we experience is a very small part of the totality of existing reality.* In some sense it might be said that we have here continued the process begun by Copernicus; not only are we not central in the universe, we are not the whole of the wavefunction! There is no doubt that this sort of interpretation of quantum theory offers hope for understanding at least some of the coincidences that are apparently necessary for life to be possible (recall the discusson of the Anthropic Principle in section 4.6). Notice here that the unreduced wavefunction does not contain only parts where particles have different spins, it also contains parts where there are perhaps no particles at all, or perhaps no galaxies, parts where much of physics and cosmology are very different to what we observe. Just as we live in the particular region of the galaxy where life is possible (i.e. on a suitable satellite of a suitable sun), so we live in a particular "part" of the wavefunction. Other parts, where maybe the "vacuum" values are different, resulting in different values of the constants of nature, may well not be suitable for life. Indeed, presumably most of the wavefunction is not suitable for life, and it could be that we exist only where the wavefunction is very tiny. If some outside observer could make a measurement, in the sense of orthodox quantum theory, the chance of his finding a universe with life in it would be fantastically small. This is no more unacceptable than the fact that most planets are not suitable places for life to develop. Just as there are a lot of planets, and we live on one that *is* suitable, so there are a lot of possibilities inherent in the wavefunction, some of which allow us to exist.

There is, however, a serious problem with this model as we have described it so far. We know that the statistical predictions of quantum theory are correct in the sense that $|\alpha|^2$ and $|\beta|^2$ are the probabilities for me to observe $+$ and $-$. But what can this mean in a model where in fact I observe **both**: one observation corresponding to a Me in one world and one to a Me in the other? We have been very careful to say that, in this interpretation, everything changes only in accordance with the Schrödinger equation, so nothing else ever happens. Of what then can $|\alpha|^2$ (more generally $|\psi|^2$) be the probability? We have removed the problem of what causes measurements by the expedient of saying that they do not occur. In so doing we have very truly "thrown out the baby with the bathwater". I believe that this is a decisive objection to the version of the many-worlds idea presented so far in this section.

Can we save the situation? That is, can we keep the advantages of this type of interpretation, which we have noted above, and still retain a satisfactory theory? In a recent paper (Squires, 1988, 1990), which I entitled *The unique world of the Everett version of quantum theory*, I suggested one possibility. To explain this, we begin by calling the world described by wavefunctions changing only in accordance with the Schrödinger equation, i.e. as in Everett, the world of (quantum) physics. Then we suppose that, in addition to the world of physics, there are things we might call **selectors**. These have the power to select results for particular observations (eigenvalues of particular observables). Thus a selector, in a world described by the wavefunction of equation 11.14, might select either the $+$ or the $-$ state. If we suppose that the selections are made at random, then the only weight function that is readily available to define the measure, relative to which they are random, is $|\psi|^2$, so the selector will automatically select a result in accordance with the probability rule of quantum theory. This is in agreement with our experience of the result, so it is reasonable to assume that it is a selector that tells our consciousness what result to "see".

Note that there is an important distinction between the classical measuring apparatus, which is implicit in the orthodox interpretation (section 11.3), and the selector we have introduced here. This is because the latter is not assumed to reduce the wavefunction, or indeed to change it in any way. Thus it does not affect what we have defined to be the world of physics.

The best analogy that I am aware of for this model was given to me by an undergraduate, Amanda Zapowski, at the Schrödinger Centenary Conference in London. We imagine a television set which has a large number of buttons for choosing programmes (see figure 11.2). A number P_1 of these are tuned to channel 1, a number P_2 to channel 2, etc, where we have introduced a suitable numeric title for the channels. The selector in this analogy is the person who switches the set to a particular channel. We must suppose that he does this in the dark, so the probability of obtaining channel i will be proportional to P_i. The world of physics here corresponds to the channels being transmitted and to the television set, and of course they are not altered by the operation of the selector.

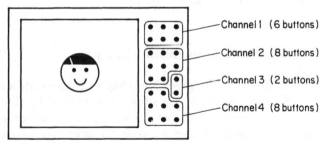

Channel 1 (6 buttons)
Channel 2 (8 buttons)
Channel 3 (2 buttons)
Channel 4 (8 buttons)

Figure 11.2. Illustrating random selection of a television programme. The probability of getting a particular channel is proportional to the number of buttons tuned to that channel, e.g. 1 in 4 for channel 1, etc.

Have we gained anything by adding this new concept? I believe that it is the only way we can make any sense out of the many-worlds idea, such that its advantages as listed above are maintained. The price we have paid for this, namely, the introduction of something new, unexplained, and, by definition, outside physics, may seem too high. However, as we shall discuss in more detail in the next chapter, it is now an obvious suggestion that selectors are, or are closely related to, conscious minds. In this way they are not something new, but rather things we know exist. However, as we shall see, new problems arise.

For completeness, we now note an alternative suggestion for giving meaning to the quantum probabilities. This is to assume that, on measurement, the world of equation 11.4 (for example) really does "split" into N_α worlds with + spin and N_β worlds with − spin, where N_α and N_β are integers which satisfy

$$\frac{N_\alpha}{N_\beta} = \frac{|\alpha|^2}{|\beta|^2}.$$ 11.16

Thus the number of worlds with + spin, for example, is larger when $|\alpha|^2$ is larger.

A model of this type reintroduces all the problems which the many-worlds idea was meant to solve: When does the split happen, What can cause it, etc? It also has problems of its own: What can it mean for the world to split, in what "space" are the parts supposed to separate, and, finally, what do we do if the ratio on the right-hand side of equation 11.16 is irrational? I cannot regard this suggestion as a serious possibility.

11.7   Knowledge and quantum theory

The considerations of the preceding chapter suggest an amusing paradox. We consider again the situation described by the wavefunction in equation 11.14. As we already saw, if "you" ask me the result of the measurement then we reach the situation described by equation 11.15, which is, to some extent, quite acceptable. Suppose, however, that instead you just ask me whether I know a definite result for the measurement. (I learned from David Albert that this is an interesting question.) It is a simple consequence of the linear laws of quantum theory that I will answer "yes" (because this is the answer I would give, correctly, in either of the two parts of the wavefunction considered separately). The wavefunction will have the form

$$|\Psi> = [\alpha|+, Me^+ > + \beta|-, Me^- >] \, |I \, know \, a \, result >,$$ 11.17

where the notation has been simplified. Clearly I have lied; I am in a mixture of two states, but I have asserted that I am in one!

Everything in this argument depends on assuming that there is nothing outside physics, and that physics follows the linear equations of quantum theory. But physics cannot lie; the physical world cannot be inconsistent with itself. What has gone wrong?

The *physics* described by the above equations is surely correct, at least for microscopic systems. Provided we replace Me by a microscopic object, then few would dispute that equation 11.7 gives the true wavefunction. Indeed, there are experiments which confirm this.

The paradox has arisen because we have inserted into physics the concept of "knowing", i.e. we have tried to give meaning to some of the readings. I wonder whether this is another indication that physical systems do not "know" anything, and that in themselves they do not have any meaning. (Recall similar comments in sections 3.3 and 9.4.) Such a concept as "knowing" is not a part of physics, so perhaps we should not be surprised that when we put it into a physical equation we find a paradox. The linearity of quantum theory cannot be reconciled with the experiences of the conscious mind. Whether the non-linearities that seem to be required for measurements are in any way associated with the conscious mind, and, if so, how we can show this, are problems that, of course, we cannot (yet) solve.

In this chapter we have looked briefly at what appear to be the most promising ways of understanding the reality behind the quantum world. The fact that the suggestions seem to be rather desperate is an indication that the problem is not an easy one to solve. We live in a very strange world. That it contains such amazing macroscopic objects as people, should perhaps not be surprising to anybody who has seriously tried to understand an electron!

Chapter 12

Conscious mind
and quantum physics

We shall now look again at some of the problems of conscious
mind in the light of what we have learned about quantum theory.
We shall follow certain rather wild speculations. Since we have
so little to guide us we shall almost certainly go in the wrong
direction, perhaps by being too wild, but more probably because
our imagination is too limited to create an adequate picture of the
reality we are trying to describe. Similar, or related, ideas to those
of this chapter may be found in Wigner (1962), de Beauregard
(1974), Walker (1979), Mattuck (1984), Margenau (1984), Eccles
(1986), Fröhlich (1987), Squires (1988, 1990), Stapp (1989 a, b)
and, I am sure, in many other places.

12.1 Why does consciousness enter physics?

The previous chapter was intended to be about physics, in particular about quantum physics (there probably is no other sort of
physics except as an approximation). In the course of the discussion we met the idea of conscious mind several times; it seemed
to arise naturally. This is a rather remarkable fact. Certainly it
would not normally happen in a discussion of classical physics. It
is therefore important that we try to see why it happened. Why,
in trying to understand the behaviour of the microscopic world,
did we apparently need the idea of conscious mind? I will try to
suggest an answer to this question, though I am not sure that I
really know what the proper answer is.

As we noted in section 5.2, our knowledge of *everything* comes
from our conscious mind, so it is obvious that when we compare

any theory with experiment we are really comparing it with the experience of conscious mind. In this sense, conscious mind is essential even in classical physics. However, classical physics is *about* the things that are experienced (indirectly) by the conscious mind. We are aware of particles and their positions, and these are the things which enter, for example, into Newton's laws of motion. Thus, in describing the theory, I can conveniently forget about the means of observation, i.e. about conscious mind. A picture something like that of figure 12.1(a) describes this situation.

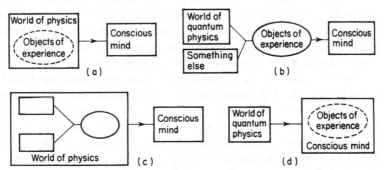

Figure 12.1. Various ways in which conscious mind might relate to experiences of the physical world.

In quantum physics the entities which appear in the theory are not those of conscious experience. The theory talks about wavefunctions, which really have remarkably little to do with the observable quantities. This is very obvious of course in a hidden-variable interpretation. More generally, we can see it from the fact that any nice smooth wavefunction is likely to be non-zero essentially everywhere, so all positions of a particle are possible. Quantum theory is a very good, deterministic, theory of wavefunctions, but it does not talk about the things we experience. The wavefunction will tell us the probability of experiencing particular values for the observables, but "something else" is required to create these values. Indeed quantum theory itself does not tell us which quantities will be the ones we actually observe (Squires, 1989c). A picture more like that of figure 12.1(b) is therefore appropriate. This picture shows that whereas classical physics tells us about the world as it is, quantum theory tells us only about what will happen when we observe the world.

Now it is not hard to make the first two pictures in figure 12.1 look the same. We merely add the extra "box" so that figure 12.1(b) becomes figure 12.1(c). Some of the interpretations we suggested in the last section, e.g. that of GRW in section 11.3, follow this line. It has many features with which we are uncomfortable and which we do not expect to find within physics. There is non-locality and the need to introduce some random input. Also the new things we add to the quantum physics box of figure 12.1(b) must not affect the statistical predictions, which are already correct. In other words, even though the physics of figure 12.1(b) does not contain the quantities of experience, it makes correct predictions about them. These predictions are not necessarily incomplete until we reach the level of conscious mind, and the discrepancy then is such that it is not unreasonable to suppose that we can locate it as "late" as possible, and thus assume that conscious mind itself is the cause, i.e. that conscious mind actually creates the objects of its experience. We would therefore draw a picture more like that in figure 12.1(d). The objects of experience are then a part of conscious mind and have no existence without it. Note that it would seem very unreasonable to do a similar thing in a classical situation. If we changed Newton's laws so that we did not obtain the results of experience, we would not be happy to say that the new laws really did correctly describe the positions of particles, but that our experience misled us into believing they were at different positions.

All this is very vague, and there may well be better answers to the question. What is certainly true is that, for some reason, in discussions of the measurement problem of quantum theory, very similar questions are asked to those which occur in discussions of consciousness: Are macroscopic systems responsible; is it something to do with complexity or is something *new* required; can this something new be regarded as a part of physics or not; can cats do it or only humans? The remarks in section 5.8, that a physical mechanism for consciousness is likely to imply some type of "pansychism", apply again to any physical mechanism which allows a system to make a quantum observation. For example, if there are terms in the Schrödinger equation that cause rapid wave-function reduction when macroscopic apparatus is involved, as in the models of section 11.3, they will have a small effect even for microscopic systems. In contrast to the case of consciousness, such small effects are in principle observable. Whitehead's "sporadic flashes" of mentality (section 5.9) might have small observable consequences.

At the very least we can say that we have met two things, conscious mind and the "something" that can make observations, which we do not understand and which seem not to be in present physics. It is therefore worth exploring the possibility that they are in some way related to each other.

In the remainder of this chapter we will look in more detail at one explicit way in which conscious mind might occur within quantum physics, namely, that suggested in section 11.6, where conscious mind might be the "selector" introduced in the many-worlds interpretation. Related suggestions are that of Wigner, already discussed in section 11.4, where conscious mind actually reduces the wavefunction, and that of Stapp (1989b) in which a conscious experience is in some way related to a collapse of a brain wavefunction. We shall however work in a many-worlds situation, where collapse does not occur.

12.2   The unique world of conscious mind

We return again to the quantum theory wavefunction for a two-state system in which I have observed the outcome of an experiment to measure a spin projection (equation 11.14). We write this in the simpler form:

$$|\Psi> = \alpha|+, Me^+> + \beta|-, Me^->, \qquad 12.1$$

where we have omitted to refer to the states of whatever pieces of apparatus might be involved in making the measurement. This wavefunction is the complete description of the quantum world. As we have already noted, however, it does not properly describe our experience because it contains two results, not one.

We therefore suppose that, in addition to the quantum world, there is *my consciousness*. This selects one or the other outcome of the experiment. (In general, this is how I define a measuring device. It is an apparatus that sends signals to my brain so that brain states become correlated with eigenstates of whatever I wish to measure, as in equation 12.1. Then I can select a result for the measurement.) Normally my consciousness will select in a random way, with the probability weights determined by the values of $|\alpha|^2$ and $|\beta|^2$.

Thus, in this sense, one part of the wavefunction has become "more real" than the other. It is the part, the *only* part, of which I am aware. We recall that it seemed to be necessary to refute any suggestion that I could become two *Me*'s, each aware of a different result, because this would not allow the possibility of getting the statistical predictions of quantum theory. The present idea also seems preferable on aesthetic grounds. The *physics* of course has not changed by my being aware of only one result. This is because we were careful to leave my consciousness outside of what we chose to call "physics".

If we wish to describe what is happening here using the many-worlds language, we would say that the world separates into two branches. In what I think is the original version of the model, "I", that is my conscious mind, would go into both. Instead we are now claiming that "I" go into one or the other branch, selected probabilistically according to the relative magnitudes of the appropriate components of the wavefunction.

There are clear similarities between this model and hidden-variable theories. In the latter there is also the sense in which one branch is special, because that is where the actual particle is. However, the present model is more idealistic (recall section 5.2); it does not have trajectories for the particles; indeed the external world does not even have particles; these are entirely a creation of the conscious mind; like free-will and redness, they are experiences.

We now come to the major problem with this model; a problem which becomes immediately evident when we ask what happens if, following my observation, *you* make a measurement of the same quantity. Normally, it would be expected that you will obtain the same result. Indeed, if you do not, we would attribute the difference to faulty experimental techniques, or to the fact that one of us is misreading the data; we therefore ignore this possibility. The question then is: *How does the theory ensure that you will obtain the same result?*

There is no problem here if you make the measurement in the simplest possible way, which is by asking me. However, you could choose to make your measurement well away from me and from my influence. To be quite specific, we could suppose that the *up* and *down* paths when the particle has passed through the Stern–Gerlach fields are well separated, as in figure 12.2. Then, if I made the measurement by observing whether the particle passed through a detector placed to detect particles on the *up* path, you could similarly use a detector in the *down* path. Clearly if my observation is that *I have seen the particle* (which means that I have become aware of the *up* path, i.e. the first term in the wavefunction of equation 12.1), then you must not see it. The probability of your seeing it must be zero, and not $|\beta|^2$ as given by the usual rule of quantum theory.

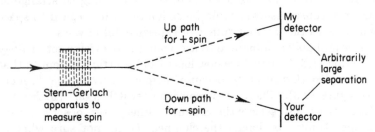

Figure 12.2. Illustrating two observers of the same process. They must obtain the same result, regardless of the separation.

This then is our problem: how is the wavefunction going to tell you that there is no particle where you are looking? Alternatively, if the wavefunction does not tell you, what does? These are difficult questions, but before we try to suggest an answer we should note that they are not new, and they are not unique to the particular model we are discussing. All solutions are unattractive; recall, for example, the difficulties in wavefunction reduction and the problems with the EPR type of experiment, which are all aspects of this same problem.

Unless we abandon the claim that my observation does not alter the physics, i.e. the wavefunction, then it is clear that there is no answer to the first problem of the above paragraph. This is a reflection of the fact that our physics is "local" (the Schrödinger equation is a very respectable, classical, equation), whereas we know that something has to be non-local. It is with considerable hesitation that I suggest that the answer must lie in some sort of "universal" nature of consciousness. Many physicists on reading such a remark will wish to go no further; others, mainly non-physicists, will probably welcome the introduction of such a concept.

What I am suggesting here is that when I have made my observation, then **Consciousness** has decided where it goes. Here I have used a capital "C", because I am not referring to the consciousness of an individual but to something else that presumably contains individual consciousnesses. The non-locality, which we know is a part of total reality, is now not contained in physics, but in the universal consciousness. If we are willing to accept that such a thing exists, and we discuss this issue further in section 12.5, then it is a more natural place for the required non-locality, because there is no reason to think that something so strange as universal consciousness should be as limited, with regard to space and time, as the familiar things of experimental physics.

It is probably helpful here to use again the television set analogy of section 11.6. The picture now has to look a little different to that given earlier (figure 11.2), because there can only be one "tuner" (see figure 12.3). The first person who "switches on" (makes the observation), also selects the channel. Other, later, observers have no opportunity to change the channel. (I am not sure whether this is the method by which possible family conflicts are avoided in the actual world of real television sets, but I would guess that it is not.) The fact that people generally experience the same world is therefore due to the fact that the human family has only one television set.

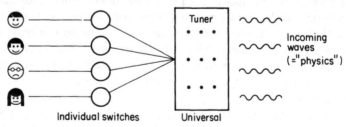

Figure 12.3. Each observer can decide whether to watch, i.e. switch on, but there is only one tuner, so all see the same programme.

Although the suggestion we are making here does not fit naturally into the way physicists normally perceive the world, and although it is true to say that we have not really solved the problem but have simply lost it in a bigger one, I do not think that there are any consistent alternatives which keep the advantages of the many-worlds idea. If we instead pretend that the first observation somehow "tags" the wavefunction, so that future observers will know which piece to select, then we might just as well say that it completely reduces it. This then becomes the model of section 11.4.

Before proceeding, we must consider one possible objection to all theories of this type, an objection that was already hinted at in section 11.4: before conscious mind came into the world, there were no particles, nothing had ever decayed, even the vacuum state of the world had not been fixed, so the parameters of the universe did not have specific values; the universe as we know it, with galaxies, and suns and moons, with hydrogen and helium, with cosmic rays and micro-wave backgrounds, etc, just did not exist. Unless we are to invoke a "divine" conscious observer, this was the state of affairs until men (or animals?) came into being. Although all this seems very counter-intuitive, it becomes less so when we realise that the reason why these things did not exist was because they were lost in a much "bigger" world. The reality that is "out-there" is a wavefunction with a richness that conscious mind can never experience. Conscious mind "creates" particles and such like because it has a limited vision of all that is. For some reason it has to select a tiny part of the wavefunction, a part that corresponds to things like near-eigenstates of position.

The remarks in the above paragraph do not mean that we cannot regard our world as having had a well defined history. Although the wavefunction at times prior to human observation contained many parts (i.e. terms in some expansion in a complete set of states), none of which were at the time in any way special, only some of them have contributed to the world experienced by conscious mind. It is these that define the history of our world. A grossly oversimplified example is given in figure 12.4.

We shall now describe three, highly speculative, suggestions that seem to occur naturally in the context of the model we are considering. They are the topics of the next three sections. Further discussion of possible implications of this model are given in the final chapter.

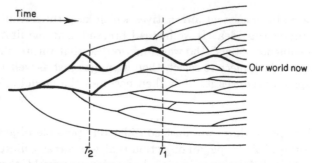

Figure 12.4. Showing how our world has a specific history. The branches represent various terms of an expansion in a complete set. Only the branches denoted by the heavy line contribute to our world, and so define our history. It is unique back to the time T_1. At time T_2 there are three possibilities e.g. three different electron paths which later come together to interfere.

12.3   Free-will and quantum physics

In chapter 7 we stated our view that free-will is not in any sense incompatible with determinism. We also suggested (section 7.5) that the experience of free-will might occur when some "part" of a person exercises a controlling influence upon some other part. Of course this is not generally the case: for example, within our body the secretions from various glands control other bodily processes, but we are totally unaware of this; none of it passes through our conscious mind so we have no experience of exercising free-will here. The question that arises is whether there is some *natural* way of splitting a person into two parts, as in figure 12.5, so that the control of one by the other corresponds to the experience of free-will.

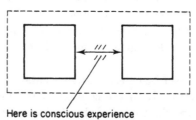

Figure 12.5. Is there a natural split of a person so that the controlling influence of one part on the other gives the experience of free-will?

In the light of the last section, a possible such split is that between the part of a person which is described by quantum physics, and the conscious mind, which is here defined in terms of its role for making observations, i.e. as the selector of section 11.6. Can this split be utilised as the origin of the sensation of free-will? At first sight the answer to this question is that it cannot, since the role of the selector is merely to make some branch of the wavefunction into the experienced world according to the probabilistic laws of quantum theory; in no sense does it exercise any control. All this however depends upon the selector making a random choice. *Why can it not be possible that in some cases the choice is not random but deliberate?* Is the choice normally random simply because we do not *try* to choose deliberately? In the television analogy of section 11.6, the choice was random because we made it in the dark. We could however have put the light on, and then, given some information about the set and the available programmes, we could have *chosen* what we see.

We are here suggesting the following model for the experience of free-will. A particular choice of action is to be determined. The process of realising that there are possible choices, and thinking about their relative merits, will have set up in the brain a quantum state described by a wavefunction. This will have the property that the action which occurs will depend upon which particular part of the wavefunction (in the sense of expansions like those in equation 12.1) becomes the experienced reality. Lack of conscious choice means that this happens statistically; conscious choice means that the selector actually determines the outcome, rather than leaving it to chance. It is the act of "observation" of a particular brain state that makes the corresponding action happen in the world of which we are aware. Note that in this picture, far from free-will being associated with randomness, i.e. with lack of determinism, it is associated with the replacing of the normal indeterminism of quantum theory by deliberate choice.

For later use, it is worth expressing these ideas in more mathematical language. We shall use the Greek letter Φ to describe the brain wavefunction and we write this as

$$|\Phi> = |\phi_1> + |\phi_2> + |\phi_3> + \text{etc}, \qquad 12.2$$

where the states on the right are such that if the brain is in $|\phi_1>$, then a particular action, say *action* (1) will occur, and similarly for the other states. Then, when the person's consciousness observes a particular $|\phi_i>$, the corresponding action (i) will be that which occurs in the observed world.

Although I have here tried to describe this suggestion within the context of a particular understanding of the many-worlds interpretation of quantum theory, it could work equally well if some sort of consciousness-related wavefunction reduction is involved as in section 11.4. The general idea could be modified in other ways. Eccles (1986), for example, suggests that the effect of the mental intention might be to change the probability of synaptic emission, rather than force a particular choice. Then *"the reliability of mental intention is derived from the integration of the chance happenings at the multitude of presynaptic vesicular grids on that neuron"*.

There are no obvious experimental implications of the idea of this section, and so no ways in which it can be tested (see section 10.7, however). It fits in with the fact that our experience of free-will suggests that there is a continuity between the two extremes of, on the one hand, definite conscious choice and, on the other, just letting things happen. Referring again to the issue of moving through time (see end of section 8.5), we experience this very much as reading a book, where we have some options of actually deciding upon the plot ourselves! Whether it is reasonable to expect that the brain contains genuine *quantum effects* as we seem to be requiring, or whether it should be expected to behave in an essentially classical way, like a macroscopic system, is something we discuss further in section 12.6. First, we see what happens if we try to extend the idea of this section to processes outside of the brain.

12.4   Psychokinesis and quantum physics

It is important to realise that, in the model discussed in the previous section, the "observation" that was made, and that gave rise to the particular action, was purely an observation of one of a set of brain states. These brain states were not correlated to any states of systems outside of the brain. Thus the observation did not determine anything about the outside world, and was therefore very different to the sort of thing we discussed in section 12.2. The difference can be seen clearly by comparison of equation 12.2 with, for example, equation 12.1. If we wish to put the state of the outside world, described say by $|\Psi>$, into equation 12.2, we would put it in as a "common factor":

$$|\Phi, \Psi >= \{|\phi_1 > +|\phi_2 > +|\phi_3 > +\text{etc}\} \; |\Psi > . \qquad 12.3$$

Here, in contrast to equation 12.1, there is no correlation between the brain states and the states of the outside world. The effect of the conscious choice on the outside world then arises only from the actions of the person's body, according to the usual laws of physics.

We could however try the possibility of extending the ideas of the above section to allow a person's conscious mind to make a deliberate selection, rather than a random one, even in cases of correlated wavefunctions like that of equation 12.1. This has drastic consequences! It means that when I observe whether a particle has + spin or − spin, *I can choose what I will see*, and hence that I can decide what the spin of the particle actually *is* (in the sense that this is determined by observation as in section 12.2). This is a form of (quantum) psychokinesis.

It is clear that this extension of the ideas of the last section has exciting and testable consequences. Is it a valid extension? Without having any "mechanism" for the process of selection it is hard to see how we can answer such a question. One obvious distinction between the two processes is that the brain states observed by conscious mind in the act of making a decision, as in the previous section, were assumed to be only microscopically different. The states that arise from a measurement, however, will normally differ macroscopically. This is true even if the measurement concerns a single particle. For example, if the measurement is made by a photographic plate, then the two results correspond to different tracks, which in turn give images at macroscopically different points on the retina. This could mean that conscious deliberate choice is possible for brain states, but not for the results of an external measurement.

Alternatively, we might want to argue that, since conscious mind is not spatially restricted in the way that "physical" objects are, it could make the deliberate choice at the level of the actual quantum event, i.e. in the wavefunction of the particle before, or independent of, the actual information reaching the brain. This would suggest that quantum psychokinesis would be possible. It would seem, however, to be very difficult, because once we abandon locality, it is hard to see how we could know *which* electron we were trying to influence; it could be one on the moon or the one in our apparatus (cf. remarks in section 6.6)! Regardless of the mechanism involved there is one important additional difference between the two types of choice. This lies in their very different evolutionary status. It is obviously very advantageous to a species to be able to choose its own behaviour (perhaps this is not quite so obvious; recall remarks in section 3.4.). On the other hand, there does not seem to be any reason why the sort of quantum influences on the outside world that we are discussing would give any advantages (particularly to species that did not know about quantum theory). Thus, even if we *could*, in principle, perform quantum psychokinesis, it is quite likely that we would not know *how* to do it.

Nevertheless, the potential might be present, and it is surely worth looking for the possibility. In section 6.4 we discussed the presently available evidence.

12.5   Universal consciousness and clairvoyance

In section 12.2 we reached the disturbing conclusion that the many-worlds interpretation, as we have understood it, seems to require that information reaching *my* consciousness must also reach some sort of "universal consciousness", which can then convey some information to *your* consciousness. Such an idea, though unnaceptable within physics, would meet almost unanimous support elsewhere. (This of course says little about its validity!) The extreme form of the idea would assert that *Consciousness is one*, i.e. there is only one conscious mind. I turn to the physicist whose name appears most in this book for a simple statement of this view, and for its justification:

> *The reason why our sentient, percipient and thinking ego is met nowhere within our scientific world picture can easily be indicated in seven words: because it is itself that world picture. It is identical with the whole and therefore cannot be contained in it as a part of it. But, of course, here we knock against the arithmetical paradox; there appears to be*

*a great multitude of these conscious egos, the world however
is only one .... There is obviously only one (way out) ...
namely the unification of minds or consciousnesses. Their
multiplicity is only apparent, in truth there is only one mind*
(Schrödinger, 1958).

With reluctance, I have to say that I find this statement hard
to accept. The assertion that the multiplicity is only apparent, is
surely self-defeating. Schrödinger is operating within a very idealistic framework (conscious mind *is* the world picture), so reality
surely is that which is apparent, and *there is nothing more apparent than that my mind is distinct from yours.* As we saw earlier
(section 3.1) *privacy* is a key feature of conscious mind, and this
fact is probably one of the reasons why it is so difficult to bring
conscious mind into the realm of physics.

In one sense of course the philosophy of idealism might be said
to require mind to be one. Otherwise it has no defence against
the argument that the commonality of the experience of different
minds can only be explained by their being the experience of one
real external world. Something has to be unified, so, if there is
no external world, it must be the mind. If the apparently many
experiences of a tree are really only one experience, then we do
not need a real tree to explain it. Arguing this way, however,
we seem to meet the devastating contradiction noted above. One
experience dominates all others: my consciousness of myself. I can
try very hard to be a "caring", moral, person, and to feel the hurt
and pain of somebody else (we return to this below), but I do not,
and cannot, feel it like my own. The little girl who was praised
by her mother for being the only one who did not laugh when a
girl at a birthday party fell and hurt herself, was indeed accepting
praise unfairly if she was the girl who fell! We may not like this
fact; we may see it as a sort of moral defect of humanity, but we
surely cannot deny that in the world as it is, it is true.

Schrödinger (1958) also gives as evidence for the oneness of mind
the fact that we never experience plurality of consciousness: *The
doctrine of identity can claim that it is clinched by the empirical
fact that consciousness is never experienced in the plural, only in
the singular. The fact is of course correct. My* consciousness is
certainly *one* thing. As we saw in section 3.4 this is crucial in
making me *me*, and in providing for continuity of experience. Much
of the remarkable power of my brain to synthesise information may

be related to the oneness of my consciousness. However, all this is about *my consciousness*. It says nothing about another's. Indeed the point can be used the opposite way. The fact that I only ever experience consciousness in the singular demonstrates how insular, or private, my consciousness is. It does not tell me that my consciousness is the same as yours.

What then, if anything, can we mean by universal consciousness? What is it that gives rise to the feeling that there is a unity in the world of consciousness, a feeling for which, again according to Schrödinger (1958), there is *miraculous agreement between humans of different race, different religion, knowing nothing about each other's existence, separated by centuries and millennia, and by the greatest distances that there are on our globe?*

One picture, which sometimes is implied, would appear to regard consciousness as a space, e.g. a space of two dimensions, in which individuals can move. At each point of this space there is a "degree of consciousness". If we could relate this to a single number then it might correspond to "height" above the two-dimensional plane. It is hard to see in such a picture what the "things" that move about actually are. They are not physical bodies, and cannot be individual consciousnesses since these are not supposed to exist. Equally if my position in the space just refers to the state of my mind, I still need the concept of "my mind", or at least of something that is *mine*.

A better picture might be to regard my consciousness as a *part* of universal consciousness. Then it may only be a question of semantics whether I say that there is one consciousness with many parts, or that the parts are individual consciousnesses which collectively make up what I call universal consciousness. Within such a picture it may be sensible to ask questions like: *Is a new consciousness created everytime a (conscious) brain is formed, or does each new brain "capture" a little piece of consciousness?*

One thing is certainly true: *direct* communication between your conscious mind and mine, if it exists at all, is very hidden and obscure. It is certainly more difficult than indirect communication through the medium of the physical world. It is in this context that the suggestion in section 12.2, which arises from one way of understanding quantum theory, has to be seen.

Let us look in a little more detail at what we appeared to require in our model. In figure 12.3 the universal aspect lay "between" our individual consciousness and the external world (of quantum physics). Because we all had to use the same "television tuner", we had a common picture of the external reality (we all tuned in to the same channel). An alternative way of obtaining the same effect would be to use a picture more like that in figure 12.6, where the universal consciousness is further away from "the external world" than the consciousness of an individual. (I must admit that I am not very clear what these pictures mean, if anything. Certainly the spatial distribution in the diagrams has no significance.) The second picture is perhaps the one that has the most natural appeal. We might regard it as indicating three "levels", the physical world, the individual conscious mind and the universal consciousness.

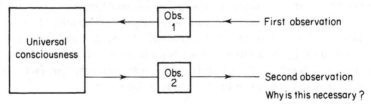

Figure 12.6. A possible relation between universal consciousness and individual minds. Why, however, is there apparently no direct communication through the universal consciousness?

The second picture, figure 12.6, however, seems to suggest that after I have established the result of some particular experiment, you could obtain it directly from universal consciousness without your having to do the experiment. If consciousness has to tell you what result to get when you do the experiment, why does it not just tell you the answer, without your having to do the experiment? This phenomenon, if it happened, would be quantum clairvoyance. Further, if we once admit the possibility of information travelling between conscious minds, without using the known physical channels, then there does not seem to be any reason why we should restrict this to information about microscopic events. This is contrary to what we might reasonably expect with psychokinesis. The evidence for clairvoyance was briefly discussed in section 6.4. To "first approximation" it does not exist, but there is some rather vague statistical evidence that, in a small and uncontrollable way, it might occur. Such evidence is hardly convincing as indicating a real effect, but seems to warrant further investigation.

Does the idea of universal consciousness give us anything else? We noted above the obvious fact that we each feel our own pain more that the pain of another person. This is why we are basically "selfish": we care more about what happens to ourselves than we do about what happens to somebody else. However, we do care to some degree about the pain of others. Compassion and sympathy do exist; they too are a part of the reality of life. It has been suggested that a possible origin for such feelings might lie in the fact that we actually *do* feel the pain of another human body, though only to a small degree. This would presumably be expected if the idea that there is only one universal consciousness was taken seriously. This however is not the way that the experiences of the feelings of compassion and such like, as they come to most people, *seem* to occur. We discover the pains of another person through the usual physical channels, and we recognise some sort of "moral compulsion" to try to respond *as though the pains were ours.* Even if the knowledge of another person's pain were to reach us through a (non-physical?) link via universal consciousness, the fact that the pain is not mine would still seem to be clear.

In 1956 Carl Sandburg introduced the "Family of Man" photographic exhibition in the New York Museum of Modern Art with the words:

> *There is only one man in all the world, and his name is All Men; There is only one woman in all the world, and her name is All Women; There is only one child in all the world and the child's name is All Children.*

It is part of what we are as people to respect such a statement, but we respect it not so much because we believe it to be true, but because we recognise its validity as a moral call that our conscious minds should be willing to transcend the limitations of their own unique individuality; not to deny that individuality, but properly to fulfil its function. It would be excellent if quantum theory, properly understood, could be seen to give some rational support to this moral call, and so help us to recognise that the concerns of one are the concerns of all (see the article *Quantum physics and human values* by Stapp, 1989c).

12.6 Are brain events microscopic?

In section 12.3 we suggested that free-will might be associated with conscious mind making a deliberate choice of some event, which, according to the rules of quantum theory, would otherwise be determined at random. It is in fact more likely that we would be concerned with deliberate choice of *many* such events, the outcome of which would determine a particular physical action. However this does not greatly alter the discussion. The question we want to consider in this section is whether the key processes in the brain can realistically be regarded as being "quantum" processes, or whether they are more properly regarded as "macroscopic", so that quantum effects are negligible.

One way of answering such a question would be to assert that quantum theory applies to all processes; certainly this is the case in the many-worlds interpretation that we are assuming. Then, even in a macroscopic system, there is some quantum uncertainty. For example, the wavefunction of a macroscopic particle interacting with a barrier, which classically would prevent its passage and cause it to be reflected, will contain a very tiny part corresponding to transmission through the barrier. According to the usual probability rules the chance of the particle passing through would be negligible, and effectively we would always obtain the classical result. This is why we can normally ignore quantum effects for classical systems. However, if conscious mind has the ability to choose an outcome of a brain process, *regardless of the size of the wavefunction*, then it can always choose to have the particle pass through. Since we have no idea of the mechanism which makes the choice, we cannot rule out this possibility, but it appears to be unnatural. We would expect the mechanism to be such that it is easy to choose the outcomes for which the quantum probability is large, and effectively becomes impossible to choose those for which it is very small. (If there are a million knobs on the television which are tuned to channel X, and only one tuned to channel Y, then we might expect that we are unlikely to be successful if we want the second channel.)

We suppose then that conscious choice arises in situations when there are several roughly equal quantum possibilities. To have a specific model, we keep to the barrier penetration idea, and imagine that the choice of action depends upon whether or not some "particle" passes through a barrier. The question we want to answer is whether it is reasonable that there should be a situation in which there is approximately a 50% chance of transmission. The difficulty of doing this comes from the fact that, for macroscopic systems, the transmission probability goes from being essentially zero, to essentially one, as the velocity of the particle changes by a very small amount. This, of course, is the reason why we cannot readily demonstrate quantum effects, in experiments like barrier transmission, with macroscopic objects. The quantum randomness would be masked by the effects of errors in the original velocity. We therefore want to know whether the brain processes are more properly regarded as being micro-, or macro-, processes.

To this end we can do a small calculation. We imagine a particle of mass m, hitting a barrier of width d such that the critical velocity for classical transmission is v. There will be some velocity for which the balance between transmission and reflection rates is exact, i.e. both are equal to $1/2$. We want to find how large an error is allowed in this velocity, so that neither of these rates becomes too small, say, falls below $\lambda/2$, where λ is a number of order unity, say, bigger than 0.1. This is easy to calculate, and we find that the fractional error that is allowed in the particle velocity is given by

$$\epsilon \approx \frac{1}{8}\left(\frac{\hbar\, ln(\frac{\lambda}{2})}{mvd}\right)^2.$$
     12.4

Notice that for a reasonable error to be allowed, m, v and d should be small; this is the requirement that the system should be "microscopic".

In order to see what the above equation means in practice, we must choose values for the parameters. We take m to be 10^{-17} g. This is about one third of the mass of a synaptic vesicle, which appears to be the smallest identifiable unit associated with the neural connections. Something of this mass will contain about 10^5 atoms. For d we take 10^{-6} cm, which is about an order of magnitude bigger than an atom. It is hard to see how a barrier smaller than this could occur. For v we choose 10^{-2} cm s^{-1}, which means that the time taken to cross the barrier is about 10^{-4} s. Then for $\lambda = \frac{1}{10}$ we obtain

$$\epsilon \approx 2.5 \times 10^{-4} \approx \frac{1}{40}\%.$$
     12.5

It is apparent from this very crude estimate that we are in a marginal situation. Without further details of a particular mechanism it is hard to be more precise, but at least we can say that the possibility of quantum effects is not entirely absurd. Eccles (1986) uses a different argument, based directly on the uncertainty principle, but he arrives at a similar conclusion. Indeed it is easy to see that any "model" is likely to give qualitatively the same result. Quantum effects are always proportional to $\hbar$, Planck's constant, and they are negligible when this is "small", which means small compared with the corresponding scale of the system being considered. Now $\hbar$ has the dimensions of *mass* times *length* times *velocity* (which itself is *length* divided by *time*), and the only obvious quantity that we can make with these dimensions comes from finding an appropriate mass, length and velocity, as in the above. Hence it is clear that the condition for significant quantum effects will have the form: $\hbar$ *is not much smaller than the product of some typical mass, velocity and distance.* Essentially this is the condition we obtained in the particular model used above.

Some confirmatory evidence that tunnelling plays a role in biological processes is reported on p.21 of the May (1989) issue of *Scientific American.*

Mattuck (1984) uses a very different set of criteria, based on information theory, to test for the possibility of consciousness effects on physical processes. He is concerned in addition with whether even macroscopic psychokinesis effects might be expected.

Chapter 13

Conclusions

The value of science lies in the questions it raises not in the answers it gives. (Pielmeier, 1982)

The sections in this chapter all have questions as their titles. We try to suggest some possible answers to the questions, but these answers lead to more questions. If we have discovered anything in the course of this book it is that nothing is as simple as it appears; confident answers can only be expected from those who have not thought too much about the questions.

13.1 Is there an external world?

This is one of the few issues which we did claim to have settled when we met it earlier. In section 5.2 we rejected idealism, and asserted our belief that there does exist a real external world, whose existence does not depend upon its having been observed by my conscious mind. Perhaps the most compelling reason for doing this is that we are then allowed to try to understand that external world. However, in this process of understanding, we have discovered that external reality necessarily contains features that are very hidden from our perception; it is not simply the things we observe. Indeed it may well be that the terms in which we try to describe reality, the things we most readily observe, particles and their positions, maybe even the space–time continuum in which we apparently exist, are to some extent constructs of conscious mind. They may be extracted from, or imposed upon, a reality which is very different.

Now one argument that we might have used against idealism is that if there is no external reality, why did my conscious mind go to the trouble of inventing one? Why should the idea of a physical world ever arise? A similar problem is apparent here. If physical reality does not actually contain things like particles and their trajectories, why does conscious mind insist on describing things in these sort of terms? An explicit example of this concerns the point noted earlier that *position* seems to play a privileged role as an observable. In quantum theory there does not seem to be any reason for this; for example, momentum and position are so-called "conjugate" variables and both occur in the theory in the same way (in the caption to figure 10.6 we saw that a wavefunction could be expressed as either a function of position or a function of momentum). However, we were happy to accept the claim of section 11.3, that wavefunction collapse to *position* would be an adequate description of experience. Why is this? Is the special role of position somehow already in quantum theory, is it a consequence of the initial state of the universe, or of the particular conditions in our universe at the present time, does it come from conscious mind, or does its origin lie somewhere beyond our current imagination? (see Squires, 1989c, for further comments).

A possible way of approaching questions of this type relies on the fact that conscious mind is itself made from the observed objects of the external world, either exclusively, as in materialism (see below), or at least in the sense that it requires such objects to interact (through brains, etc) with that world. Then we could say that conscious mind needs the things it constructs in order that it can exist; alternatively, that conscious mind perceives a particular, restricted, version of reality containing the things from which it itself is constructed. Einstein expressed his amazement at the fact that we could, with some success, comprehend the external world. Maybe the truth is rather that we comprehend very little of it. Reality is way beyond our understanding, but we can understand those parts of it that are immediate to us, because *they* are what we consist of.

Our answer, then, to the question of whether there exists an external world is still a very definite "yes". However, we have only a very vague idea of what that external world really *is*. Perhaps the only thing of which we can be confident is that we do not understand it!

13.2   Is there anything else?

We return to what is often regarded as one of the central issues of philosophy: Is physics everything, or is there "something else"? As we saw in chapter 5, it is the answer given to this question that separates materialists from non-materialists. However, before we try to answer a question, we must be careful that we know what it means. We have already suggested that attempts to make the question precise are likely to be such as to make the answer trivially obvious, which would imply that the issue is spurious. In view of the long history of the controversy, this is a surprising suggestion which we need to examine more closely.

First, I would like to assure readers that we are not concerned here with the problem of finding *rigorous* definitions. Strictly speaking it is probably not possible to give a proper definition of anything relevant to the real world. Readers of this book will have already realised that we have been happy to accept intuitive ideas of what words mean (some might have been very critical of our looseness in this regard). It is within such a framework that there are problems in making the materialist *versus* non-materialist issue precise.

We try the following way of putting the question: *Do the laws of physics give a complete explanation of all that was, that is, and that ever shall be?* (Compare section 5.4.)

If we interpret the laws of physics as being those at present known, then clearly the answer is that they do not. We recall that in section 4.6 we saw that there are even some things which obviously belong to physics that are not yet understood. However, this is being a little unfair; it is possible to imagine that some small extensions of contemporary ideas, involving string theories, inflation, etc, could overcome these problems. We would then be left with the vague feeling that there are some difficulties regarding time, and also with the real problems of quantum theory, which we have discussed at length in the last few chapters. If we restrict ourselves to theories based upon orthodox quantum theory, then again the answer to the above question is clear; in the sense that is appropriate here, quantum theory is incomplete because it does not explain the process of observation.

The next step is to propose some version of quantum theory which *is* complete, at least as far as "obviously physical" effects are concerned. If we go along the many-worlds road then, as explained in section 11.6, we have to introduce conscious mind into our physical theory. It is not clear then that we would necessarily want to say that even "physics" is materialistic; if we did, it would be a form of materialism that even a convinced dualist would be happy to accept. The issue of whether conscious mind is within physics would have been settled, because we would have deliberately introduced it into physics. It would not be derived from physics, but would be an ingredient of physics. Of course we would be free to use the word "physics" to mean just quantum physics, in which case we would have a dualistic model because conscious mind would be something separate, or we could define it to include conscious mind, in which case we would become materialists. The distinction, however, can have no significance.

Alternatively, we might try to complete quantum theory with some type of mechanism for wavefunction reduction. Even here there remains the need for a random input from "outside" the theory. The only way to remove the need for this is to adopt the hidden-variable theory, which, as we noted, is classical physics plus some peculiar quantum influences. In principle, at least, it seems as though it might be possible on these lines to define laws of physics which explain all phenomena not involving, say, human brains. We would then be in a good position to pose our question: *Would these laws explain everything, including human brains?* If the answer to this question is yes, then we would seem to be very close to a traditional materialistic position. The concepts of consciousness, free-will, colour, happiness, etc, would have to be constructed from the ingredients of the theory, much as water molecules and bicycles are so constructed. I do not know of any argument that tells me that this is not possible; it seems to be extremely unlikely, but feelings are not an argument.

To proceed, we shall suppose, first, that it is not true, i.e. to be quite explicit, that *I cannot make conscious mind out of the laws of physics that explain everything else.* Except for the fact that we might want to regard conscious mind as "different" to anything else, this situation is not new in physics. As a trivial example, when the so-called "charmed" particles were discovered in November 1974, we could not make them out of the three quarks known at the time. Our response to this was obvious: we invented a new quark. (To be quite honest here there were good theoretical reasons for believing that the charmed quark had to exist, but in 1974 very few people believed them.) When we have found something not in physics, we put it in, either explicitly, or by putting in the ingredients out of which we can construct it. In this way, essentially by definition, everything is in physics. If we allow this sort of extension then the statement of materialism is totally empty, because it would be logically impossible to deny it: everything is in physics because we define physics to include everything.

There is however something about the last paragraph that is not correct. The physical theory, without charmed quarks, could not have been *exactly* correct even for processes not involving the charmed quarks. This is because in quantum theory everything is so mixed up that even if the charmed quarks are not explicitly present, they alter the results that would be seen (through so-called "virtual states"). In other words, if charmed quarks exist, there is no physics that does not involve them. If the same thing is true of conscious mind, then it would just not be possible to have a theory that explained everything except conscious mind.

Now we try the opposite assumption, namely, that *the laws of physics, defined so that they explain everything else, also explain consciousness.* In some interpretations of the word this could be said to be a statement of materialism, even if, as suggested above, it might be to some extent an empty statement. However, we must be careful to understand what we mean by the word "explain", in particular, the difference between explaining what *can be* and what *is* (recall the remarks in section 2.3). Popper (in Popper and Eccles, 1977) makes a strong attack on materialism. Nevertheless, he writes:

> *I share with materialists ... the evolutionary hypothesis ... evolution produces minds, and human language ... human minds produce stories, ..., works of art and science. All this, so it seems, has evolved without any violation of the laws of physics .... We can only wonder that matter can thus transcend itself, by producing mind, purpose, and a world of the products of the human mind.*

We can, indeed, wonder!

Presumably here we must mean the laws of physics defined without explicit reference to conscious mind. Then, if these laws are deterministic, Popper's statement is a clear statement of strict materialism. Since he does not intend this he must be using the fact that the laws are not deterministic, i.e. he is not accepting a hidden-variable type of theory. Then the implication is that "something else" has guided the way things have developed so that conscious mind, etc, have evolved. In other words physics *permitted* the emergence of consciousness; if it had *required* it, then we have a materialist model; if it did not require it then we have a non-materialist model. This, I think, is what Popper is saying. I am not sure, however, that things are quite so simple.

In figure 13.1 I have tried to draw possible scenarios for the "creation" of human consciousness, as it exists in the world at the present time. The diagrams are very symbolic and oversimplified; in particular they are drawn using what might be a far too naive concept of the direction of time and its relation to cause and effect. In (a) we have a deterministic physical universe, with some arbitrary initial condition. The discussion of section 4.6 suggests that such a universe would be unlikely to contain conscious minds. Of course, if there are enough such universes, somehow disjoint from each other, then any particular set of parameters, however unlikely, will occur in some. In (b) we have illustrated how, with a suitable, carefully selected, starting point, conscious mind might emerge. The next diagrams all refer to non-deterministic physics. In (c) we have a model in which there is random wavefunction reduction. As in (a), it is hard to see how such a model could produce conscious mind, unless the sequence happens a very large number of times. In (d) we have a "many-worlds" scenario. It is possible that this automatically ensures that there is a big enough spread of conditions so that, somewhere (in the wavefunction), brains, and even consciousness, might emerge. All the models so far come into the category of being materialistic, except that the choice of the particular boundary condition in (b) might be regarded as an input from outside physics. Such an input is made explicit in a model like that of (e), where the "choice" of path (for example, in a model with wavefunction reduction) is determined by a deliberate act from "outside". Here we have introduced the concept of a "god", who directs the way the quantum probabilities are made into particular choices. Another possibility is given in (f). In this model, consciousness is not created out of the physical world, but is present at all times. It could interact with the physical world by determining the choices, as in model (e), ultimately permitting the creation of brains with which it can become associated in a very special way. There are of course similarities between (e) and (f), and we may want to regard the box labelled conscious mind as being, or as containing, God. We would naturally regard (f) as being a dualistic model, at the opposite end from materialism. However, we could draw an extra box around the picture as in (g). Then things are not so clear, and we again meet the question of how we have made the separation of the large box into two. What is the essential difference between the two parts that makes the separation natural? In fact, (g) now looks remarkably like (a), so we have again the problem of why the apparently unlikely set of circumstances for minds actually occurred. I somehow want to say that the conscious mind box contains purpose, but I am not sure whether this is saying that the initial conditions are carefully chosen, or whether it is saying that it is the final conditions that are in fact determining what happens. As we saw in section 7.6, the idea of purpose is that the final state determines what happens, so in some sense it is a time-reversed type of causality.

I do not know how to answer the question at the head of this section. We learn about the external world through the experiences of our conscious mind, and much of what we have learned can be understood in terms of physics. It is natural then that we should try to extend the explanations of physics into the original observation mechanism, the conscious mind. It may be that such an attempt is doomed to failure. It may be that the external world will never lead us to a physics rich enough to explain ourselves. Certainly if we limit our idea of physics to something that is simple and (almost) understood, then there is no doubt that we need something else. But whatever that something is, it is not easy to see why we should rule it out from being a part of some future physics. The question in fact is best "answered" by another question, namely:

13.3   Why does it seem to matter so much?

We have reached a strange sort of paradox. A question which has always been central to our discussion, and one over which very strong opinions are expressed, seems to have almost melted

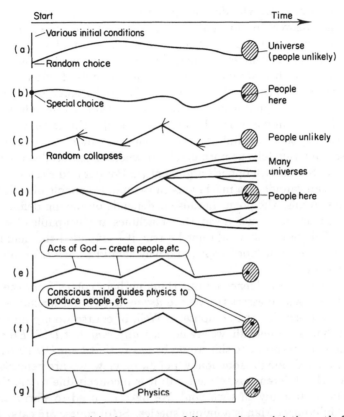

Figure 13.1. In (a) the universe follows a deterministic path from a random initial state. It is unlikely to produce people. In (b), on the other hand, the initial state is a carefully selected people-producing state. Part (c) is a non-deterministic quantum universe with random collapses. Again people seem unlikely. In (d) we have a many-worlds scenario in which all branches exist, in contrast to (c). Some might well produce people, whose observations then make their world the unique observed world. Part (e) is a quantum version of (b) in which deliberate, not random, choices are made. This creates people by design. In (f) it is assumed that consciousness is always present and reacting with the world. Again it can produce people if suitable choices are made. Part (g) is like (f) except that we have drawn a line around physics and consciousness, so it is not clear whether it really differs from (a).

away. The answer seems to be very dependent on exactly what the question means. Why does it appear to be so important?

The answer to this has already been noted in section 2.2. According to materialism, "man is a machine", and such a statement seems to deny everything that makes us what we are. But of course such a statement has no content until I know what is meant by a machine. I can either define it by some set of properties, or by saying what sort of things I want to include within it. If, for example, I use the word to describe things that are not conscious, then certainly the statement that man is a machine is false. But this tells us nothing about whether materialism is valid or not. Similarly, Popper's argument (in Popper and Eccles, 1977), that man cannot be a machine because man is capable of suffering, seems to miss the point, because it depends on having defined the word machine in such a way that machines are incapable of suffering. This of course would not be true if I extended the definition of machine to include people, and why should I not? After all it is only a word!

A powerful influence on our thinking here comes from the fact that we want to assert the we are *different* to other machines, like cars and bicycles and computers. Well, of course we are different. But that is not enough, we are not just *different*, but *DIFFERENT*. I think my response to this would be to say that I am *DIFFERENT* to you. Although from many points of view we are remarkably alike, the difference is to me total; it is inconceivable to me that I could confuse myself with you! We have no need to worry about our difference. Man is a unique species. So of course are cats. And I am not sure that there are "degrees of uniqueness".

Another major part of our reluctance to see ourselves in materialistic terms, lies in our unwillingness to recognise the significance of conscious mind. I believe we saw this already in chapter 7, where we noted a desire to ask that free-will should be more than an experience, as if there could be anything that is *more*. In the same way I think we are unwilling to see that there are many other things which are important, and real, because they are in the conscious mind. For an example of this sort of thing, I refer to a recent article of Atkins (1987) entitled *Purposeless People*. This article is an enthusiastic attempt to defend a totally materialistic view of everything. To my mind it fails in its aim (and in saying this I am not committing myself to any opinion on whether its basic claim is correct). Even the title is based on a myth; there are no

such creatures as purposeless people. Foolish, sad, unkind, envi-
ous, selfish, etc, we may too often be, but purposeless, surely very
seldom. To regard purpose as *wishful thinking and hallucination* is
to be dealing with a world which is not the one we inhabit. Atkins
can be confident that he has solved the problem, and can make remarks like: *"Consciousness ... depends on, and is the non-linear
summation of, the physical states of nerve cells"* because he has
ignored it. To be fair he later accepts that people have what he
calls a *local sense of purpose*, but he considers this to be futile
because in billions of years its effects will have disappeared. I do
not follow the argument here; many very real effects of purpose in
fact survive for much shorter times!

The world has meaning, not because I cannot describe it by
physics (that is probably a trivial matter of definition), but because
I give it meaning. There is beauty in the world, because *I see
beauty*. Like free-will and purpose, meaning, truth, beauty, hope,
love, colour, and many other things, are part of the wonder of
conscious mind. I believe that this is true regardless of whether it is
convenient to allow physics to embrace the things of consciousness,
and quite independent from whether strong AI is, or is not, a valid
possibility.

The day before I wrote the first draft of this section I read
about a seven-year-old girl in a Swiss village who was dying of
an incurable disease. She knew she was dying and had one big
wish: that she would live to experience another Christmas. As
the days passed, it became evident to her doctor that she would
die too soon. The people of the village, however, were not easily
beaten. Though they could not do anything to keep the girl alive,
they could, and they did, move Christmas! So, on some particular
day before Christmas, carols were sung, parcels were unwrapped,
Christmas food was eaten, etc. All these were physical things, i.e.
they were the motion of atoms. Now it may be true that everything
happened because these atoms moved entirely in accordance with
the laws of physics (some sort of physics); in other words, if we had
known enough about the positions of all the relevant particles in
the universe at some time hundreds of years earlier, then we could
have calculated that this would have happened. It may be true or
it may not. The question is of great interest but I do not know the
answer. There are two things I do know, however. One is that the
moment I read the story, this sometimes rather sad world became

a better place. The other is that the story is a great tribute to the subject of this book: the human, conscious, mind.

13.4   Are we responsible for our actions?

The end of the last section provides an ideal introduction to the long-ago-promised "sermon". We have repeatedly emphasised the reality of the experience of freedom. There exists something which I call "me", that exercises a degree of control over many of the things I regard as mine. To this extent, at least, I am responsible for the world in which I live. How much further my responsibilty may extend, through the non-local properties of quantum theory, through the role of conscious mind in determining the world I experience, etc, are matters for conjecture. But the reality of my freedom is not in doubt. I may not always be able to change the date of Christmas, but there is some part of this great pageant of the universe which will be, forever, as it is, because of my choice. That is a cause for celebration!

Sadly, however, the ideas of free-will and of responsibility are often discussed in a very different way. The concern is not with the reality of *my freedom*, but with whether I have the right to hold another person responsible for his actions. Put simply there is a fear that determinism takes away my right to "punish" somebody else for what I conceieve as their bad actions. Many years ago in a popular newspaper, I read an article in which it was reported that some research had suggested that criminal behaviour was associated with a particular chromosome. The article was written in such a way as to deride the research, because there was a feeling that any discovery that led to the conclusion that crime was caused by chemistry would remove the right of society to extract retribution, and that could not be tolerated (it was that sort of newspaper). Even the OCM has an article on determinism and free-will which is concerned with essentially the same problem. There seems to be an almost desperate desire to provide a justification of retribution on the basis of some concept of free-will. The argument seems to depend on the bottom line: "of course we are justified in exacting retribution from those who do wrong", so we must adjust our metaphysics accordingly. Honderich (1969) notes that for this purpose some writers try to define free-will simply as "unconstrained action", which appears to give it to a football! Smith and Jones (1986), who write very much within a deterministic, materialistic, context, introduce what they call "psychological indeterminism" to justify "our practice of holding people responsible for their actions, of blaming them for their wrong-doings... ". To my mind all such justifications are bogus.

When I refer to another person I mean the whole person, i.e. everything in the dotted box of figure 12.5. Whatever the split which may be appropriate in locating the origin of free-will (see sections 7.5 and 12.3), it now has no relevance. Part of a person, call this for the sake of the argument the person's consciousness, may well be choosing how the other part behaves, but *it did not choose what sort of a consciousness it is*. If a person is "morally weak", or just "bad", or whatever, then that may explain the actions the person takes. The person however did not choose whatever it is (was) that makes him morally weak, or bad, or whatever. A man can no more be blamed for having a weak "moral fibre", than he can for having a weak kidney. The fact that one of these generally operates through the conscious mind and the other does not is totally irrelevant here. To use our experience of free-will (the wonderful truth that there is a part of the world that is ours to create) as a pseudo-justification for some primitive desire for vengeance and retribution is to misuse it.

There is nothing very new about what we are saying here. Indeed it is basic to three (at least) of the world's major religions: Judaism, Christianity and Islam, which take the story of Genesis as being the origin of evil in the world. (That the behaviour of many of the followers of these religions reveals that they ignore this message is, sadly, all too evident.) The clear statement of the second story in Genesis is that people are "evil" not because of what they do, but because of what they are, and for that they are not responsible. More recent discussions can be found in Honderich (1969, 1988).

It is of course reasonable that societies should protect themselves against both things and people that are harmful, and this necessarily involves a legal system with penalties for certain types of behaviour. Deterrence, prevention and cure all have a part to play here. Retribution just confuses the issue, and is indeed a positive hindrance to attempts to reduce the amount of crime. We solve problems by understanding them, so we will remove crime by understanding criminals. To exact retribution is both unjust, because it adds a burden to the already burdened, and tragic, because it hinders attempts to solve the problem of "anti-social" behaviour.

13.5   What is conscious mind?

The dilemma that Descartes recognised over 300 years ago is with us still. I cannot imagine how I will ever be able to make, or even design, a machine so that it will be conscious, i.e. will be aware of its own existence, will feel happy or sad, will have a concept of truth, will care about me, or indeed will be something about which I should care. Similarly, I cannot conceive that it will ever be possible to do a calculation which will demonstrate that something has these properties, in the way that I might be able to prove that something would be a liquid, for example. This would lead me to suggest that I, and other people, and other things that have conscious minds, are completely *different* to the things that I can design and make. But then the other side of the dilemma becomes evident. I cannot imagine why, or how, I *can* be different, or what it might be that is in me, that I cannot design into a machine. Putting this in the positive way, I can readily be convinced that a machine cannot be conscious, cannot know about love, or red, etc. The problem is that the argument is *too good*. I do not know why it does not apply to me! Somewhere it has gone wrong, because *I am conscious*.

It is tempting to follow Descartes into a naive dualism, and to put a limit to the domain of physics so that it cannot include consciousness. Consciousness just *is* the things it contains: it is purpose and free-will, it is truth and love, it is red, and blue, it is joy, and it is all the many things that owe their existence to conscious mind. But all this is "poetry" and has little to do with science. Such a "thing" is not a part of the world in the way that consciousness has to be. It cannot be an accident that the colours I experience are closely correlated with the wavelengths of electromagnetic radiation.

Alternatively, I could take refuge in the comfortable excuse of complexity: the process of evolution, working over millions of years on a structure as complex as the human brain, has so strengthened the survival mechanism that it has come to seem like a *desire* to survive, and from that desire the other attributes of consciousness have developed. But this really does not say anything; it is the start of an investigation, not its result!

Naive materialism can hardly even start. A statement that everything is matter is without content since we do not know what matter is. We want to say that it is particles, quarks and leptons, etc, but what are these? In most versions of quantum physics they do not really exist, without minds to observe them.

In the early 1960's, before the quark model became widely accepted, many physicists believed in the "bootstrap" model of particles (Chew, 1966, Capra, 1975). Here, there were no really elementary particles; everything was a bound state of other bound states; they sustained each other in a, possibly unique, self-consistent, cooperation. For particles, the idea died, but maybe something similar sustains mind and matter: conscious mind requires particles which in turn only exist because of conscious mind. Perhaps the bootstrap idea of Geoffrey Chew, the process philosophy of Alfred Whitehead, and a new understanding of quantum theory will one day come together to give a coherent picture of the relation between conscious mind and the physical world.

Here, we have to acknowledge that, at present, our imaginations are inadequate, and that some new ideas are needed if we are to understand these things. It is my belief that, if this is to happen, we will realise that the problems of conscious mind and the problems of quantum theory are intimately related. When we understand one, we will have made some progress in understanding the other. There is good reason to believe that such understanding will help us to see anew that "no man is an island"; it will provide scientific support for the moral conviction that we are a part of something bigger than our individual selves (Stapp, 1989c).

Until that time we will not let the mystery of the origin of conscious mind prevent us from recognising its importance as an *ingredient* of reality. Nor will we forget that our experiences are not just a guide to the nature of the world, they are themselves a part of the world. Because meaning and purpose exist in me, they must exist in the world in which I live. However much, or little, we may learn of the nature of truth from the experience of beauty, we cannot ignore the message of John Keats' *Sylvan historian* that *Beauty is truth.*

References

Albert D (1986) in *New Techniques and Ideas in Quantum Measurement Theory*, ed. Greenberger (Academy of Sciences, New York) p.480

Aspect A and Grangier P (1987) Hyperfine Interact. **37**, 3

Atkins P (1987) in *Persons and Personality*, ed. Peacocke and Gillet (Blackwell, Oxford) pp.12–32

Bakhurst D and Dancy J (1988) Times Higher Education Suppl., **22** April, p.18

Barrow J D and Silk J (1983) *The Left Hand of Creation* (Heinemann, London)

Barrow J D and Tipler F J (1985) *The Anthropic Cosmological Principle* (OUP, Oxford)

Bartholomew D J (1988) J. R. Stat. Soc. A **151**, 1

Baumann K (1986) Il Nuovo Cimento **96B**, 21

Bell J S (1964) Physics **1**, 195

Bell J S (1987a) *Speakable and Unspeakable in Quantum Mechanics* (CUP, Cambridge)

Bell J S (1987b) in *Schrödinger* ed. Kilmister (CUP, Cambridge) pp.41–52

Benninger R J, Kendall S B and Vanderwolf C H (1974) Can. J. Psychol. **28**, 79

Blakemore C and Greenfield S (1987) *Mindwaves* (Blackwell, Oxford)

Boghossian P A and Velleman J D (1989) Mind **98**, 81

Bohm D J and Hiley B J (1984) in *Quantum, Space and Time - the Quest Continues*, ed. Barut, van der Merwe and Vigier (CUP, Cambridge) pp.77–92

Bohr N (1958) *Atomic Physics and Human Knowledge* (Wiley, New York)

Bussey P (1984) Phys. Lett. **106A**, 407
Bussey P (1986) Phys. Lett. **118A**, 377 and **120A**, 51

Capra F (1975) *The Tau of Physics* (Wildwood, London)
Carr B (1988) in *An Encyclopedia of Philosophy*, ed. Parkinson (Routledge, London) pp.76–98
Cartwright N (1983) *How the Laws of Physics Lie* (OUP, Oxford)
Chew G F (1966) *The Analytic S-Matrix* (Benjamin, New York)
Chung K L and Walsh J B (1969) Acta Math. **123**, 225
Churchland P M (1984) *Matter and Consciousness* (MIT Press, Cambridge, Mass.)
Claridge G (1987) in *Mindwaves*, ed. Blakemore and Greenfield (Blackwell, Oxford) pp.29–41
Close F (1983) *The Cosmic Onion* (Heinemann, London)
Cohen D E (1987) *Computability and Logic* (Horwood, Chichester)

Davies P C W (1982) *The Accidental Universe* (CUP, Cambridge)
Davies P C W (1983) *God and the New Physics* (Dent, London)
de Beauregard O C (1974) Found. Phys. **6**, 539
Diosi L (1989) Phys. Rev. **A40**, 165
Dirac P A M (1930) *Quantum Mechanics* (Oxford, 4th edition, 1958)

Eccles J C (1986) Proc. R. Soc. **B227**, 411
Eccles J C (1987) in *Mindwaves*, ed. Blakemore and Greenfield (Blackwell, Oxford) p.293
Edwards P (1967) *The Encyclopedia of Philosophy* (Macmillan, New York)
Englert F (1989) Phys. Lett. **B228**, 111
Erice Statement (Dirac, Kapitza, Zichichi, 1982) Etore Majorana Centre for Scientific Culture, Erice, Sicily
d'Espagnat B (1983) *In Search of Reality* (Springer, Berlin)
d'Espagnat B (1989) *Reality and the Physicist* (CUP, Cambridge)
Everett H (1957) Rev. Mod. Phys. **29**, 454
Eysenck H J (1957) *Sense and Nonsense in Psychology* (Penguin, London)

Feigl H (1967) *The "Mental" and the "Physical"* (Univ. of Minnesota Press, Minneapolis)
Fine A (1986) *The Shaky Game* (Univ. of Chicago Press)
Flew A (1984) *A Dictionary of Philosophy* (St Martin's Press, New York)

Friday J (1988) *The Guardian* **20** July, p.18

Fröhlich H (1987) in *Quantum Implications - Essays in Honour of David Bohm*, ed. Hiley and Peach (Routledge, London) pp.312–13

Gardner M (1989) Am. J. Phys. **57**, 203

Gell-Mann M (1979) in *The Nature of the Physical World*, ed. Huff and Prewett (Wiley, New York) pp.180–92

Ghirardi G C, Grassi R and Rimini A (1989a) *A Continuous Spontaneous Reduction Model Involving Gravity*, Trieste preprint

Ghirardi G C, Pearle P and Rimini A (1989b) *Markov Processes in Hilbert Space and Continuous Spontaneous Localisation*, Trieste preprint IC/89/105

Ghirardi G C, Rimini A and Weber T (1986) Phys. Rev. **D34**, 470

Gisin N (1989) Helv. Phys. Acta **62**, 363

Gödel K (1931) Monatshefte für Mattematik und Physik **38**, 173 (Engl. trans. in *On Formally Undecidable Propositions*, Basic Books, 1962)

Golding W (1959) *Free Fall* (Faber, London)

Good I J (1967) Br. J. Philos. Sci. **18**, 144

Good I J (1969) Br. J. Philos. Sci. **19**, 357

Gregory R L (1981) *Mind in Science* (Penguin, London)

Gregory R L (1987) *The Oxford Companion to the Mind* (OUP, Oxford)

Gribbin J (1984) *In Search of Schrödinger's Cat* (Wildwood, London)

Griffin D R (1986) *Physics and the Ultimate Significance of Time* (State Univ. of New York Press, New York)

Hansel C E M (1980) *ESP and Parapsychology - a Critical Reevaluation* (Prometheus, Buffalo)

Hawking S (1984) in *Relativity, Groups and Topology II*, ed. deWitt and Stora (North-Holland, Amsterdam) pp.333–80

Hawking S (1988) *A Brief History of Time* (Bantam, London)

Hodgson P E (1988) Science and Religion Forum **12**

Hofstadter D R (1979) *Gödel, Escher, Bach: An Eternal Golden Braid* (Harvester, Sussex)

Honderich T (1969) *Punishment* (Hutchinson, London)

Honderich T (1987) in *Mindwaves*, ed. Blakemore and Greenfield (Blackwell, Oxford) pp.445–58

Honderich T (1988) *A Theory of Determinism* (OUP, Oxford)

Humphrey N (1986) *The Inner Eye* (Faber, London)

Jahn R G and Dunne B J (1986) Found. Phys. **16**, 721
Jahn R G and Dunne B J (1987) *The Role of Consciousness in the Physical World* (Harcourt, Brace Jovanovich, San Diego)
Janes H (1976) *The Origin of Consciousness in the Breakdown of the Bicameral Mind* (Penguin, London)
Josephson B D (1984) in *Science and Consciousness*, ed. Cazenave (Pergamon, Oxford) pp.9–19

Kent A (1989) *Do Interactions Cause State-Vector Reduction?*, Institute for Advanced Study at Princeton preprint

Libet B, Gleason C A, Wright E W and Pearl D K (1983) Brain **106**, 623
Libet B, Wright E W, Feinstein B and Pearl D K (1979) Brain **102**, 193
Lucas J R (1961) Philosophy **36**, 112
Lucas J R (1967) Br. J. Philos. Sci. **19**, 155

MacKay D (1987) in *Mindwaves*, ed. Blakemore and Greenfield (Blackwell, Oxford) pp.5–16
Margenau H (1984) *The Miracle of Existence* (Oxbow, Connecticut)
Mattuck R D (1984) in *Science and Consciousness*, ed. Cazenave (Pergamon, Oxford) pp.49–66
Maxwell N (1988) Br. J. Philos. Sci. **39**, 1

Nagel E and Newman J R (1959) *Gödel's Proof* (RKP, London)
Nyhoff J (1988) Br. J. Philos. Sci. **39**, 81

Pagels H R (1983) *The Cosmic Code* (Joseph, London)
Pais A (1982) *Subtle is the Lord* ... (OUP, Oxford)
Parfit D (1987) in *Persons and Personality*, ed. Peacocke and Gillett (Blackwell, Oxford) pp.88–98
Parkinson G H P (1988) *An Encyclopedia of Philosophy* (Routledge, London)
Pearle P (1976) Phys. Rev. **D13**, 857
Pearle P (1988) *Combining Stochastic Dynamical State Vector Reduction with Spontaneous Localisation*, Int. Centre for Theoretical Physics, Trieste, preprint IC/88/99
Pearle P (1989) *Towards a Relativistic Theory of State Vector Reduction*, Hamilton College, Clinton, preprint

Penrose R (1987) in *Mindwaves*, ed. Blakemore and Greenfield (Blackwell, Oxford) pp.259–76

Penrose R (1989) *The Emperor's New Mind* (OUP, Oxford)

Percival I (1989) *Diffusion of Quantum States*, Queen Mary College preprint

Philippidis C, Dewdney C and Hiley B J (1979) Il Nouvo Cimento **52B**, 15

Pielmeier J (1982) *Agnes of God* (New American Library, New York)

Pippard A B (1988) Contemp. Phys. **29**, 393

Polkinghorne J (1984) *The Quantum World* (Longman, London)

Polkinghorne J (1988) *Science and Creation* (SPCK, London)

Popper K R and Eccles J C (1977) *The Self and its Brain* (Springer, Berlin)

Premack D and Woodruff G (1978) Behavioural and Brain Sciences **4**, 515

Radin D I and Nelson R D (1988) *Evidence for Consciousness Related Anomalies in Random Physical Systems*, Report from Princeton Dept of Psychology

Rae A (1986) *Quantum Theory - Illusion or Reality* (CUP, Cambridge)

Rauch H (1984) in *Open Questions in Quantum Physics*, ed. Tarozzi and van der Merwe (Reidel, Dordrecht) pp.345–76

Redhead M (1988) *Incompleteness, Non-Locality and Causality* (CUP, Cambridge)

Redondi P (1988) *Galileo, Heretic*, translated into English by R Rosenthal (Allen Lane, London)

Roland P E and Friberg L (1985) J. Neurophysiol. **53**, 1219

Romanes G J (1896) *Mind and Motion and Monism*

Rose S (1987) *Molecules and Minds* (Open University Press, Milton Keynes)

Rucker R (1987) *Mind Tools* (Houghton Mifflin, New York) (published in 1988 by Penguin, London)

Russell B (1921) *The Analysis of Mind* (Allen and Unwin, London)

Schlipp P A (1941) *The Philosophy of Alfred North Whitehead* (Tudor, New York)

Schrödinger E (1944) *What is Life?* (reprinted by CUP, Cambridge, 1967)

Schrödinger E (1958) *Mind and Matter* (reprinted by CUP, Cambridge, 1967)

Searle J (1984) *Minds, Brains and Science* (BBC, London)

Searle J (1987) in *Mindwaves*, ed. Blakemore and Greenfield (Blackwell, Oxford)

Smith P and Jones O R (1986) *The Philosophy of Mind* (CUP, Cambridge)

Squires E J (1981) Eur. J. Phys. **2**, 55

Squires E J (1985) *To Acknowledge the Wonder* (Adam Hilger, Bristol)

Squires E J (1986) *The Mystery of the Quantum World* (Adam Hilger, Bristol)

Squires E J (1987) Eur. J. Phys. **8**, 171

Squires E J (1988) Found. Phys. Lett. **1**, 13

Squires E J (1989a) Br. J. Philos. Sci. **40**, 413

Squires E J (1989b) *Wavefunction Collapse and Lorentz Invariance*, Durham preprint

Squires E J (1989c) *Why Is Position Special?*, Durham preprint to be published in Found. Phys. Lett.

Squires E J (1990) in *Quantum Theory without Reduction*, ed. Cini and Lévy-Leblond (Adam Hilger, Bristol) pp.151–60

Stapp H P (1972) Am. J. Phys. **40**, 1098

Stapp H P (1989a) *Noise-Induced Reduction of Wavepackets*, University of California at Berkeley preprint

Stapp H P (1989b) *Quantum Theory of Consciousness*, University of California at Berkeley preprint

Stapp H P (1989c) *Quantum Theory and Human Values*, University of California at Berkeley preprint (text of a talk at the UNESCO conference Science and Culture in the 21st Century - Agenda for Survival)

Swinburne R (1987) in *Persons and Personality*, ed. Peacocke and Gillett (Blackwell, Oxford) pp.33–55

Sudbery A (1986) *Quantum Mechanics and the Particles of Nature* (CUP, Cambridge)

Thagard P (1988) *Computational Philosophy of Science* (MIT Press, Cambridge, Mass.)

Thorpe J (1980) *Free Will* (Routledge and Kegan Paul, London)

Tipler F (1989) Phys. World **2**, 45

Tipton I (1988) in *Encyclopedia of Philosophy*, ed. Parkinson (Routledge, London)

Voltaire (1760) *Candide* (Engl.Trans. (1968) OUP, Oxford)

von Fraassen (1980) *The Scientific Image* (OUP, Oxford)

von Neumann J (1932) *Mathematische Grundlagen der Quanten-mechanik* (Engl. trans. *Mathematical Foundations of Quantum Mechanics*, 1955, Princeton)

Walker E H (1979) Psychoenergetic Systems **3**

Webb J (1968) Philos. Sci. **35**, 156

Weinberg S (1977) *The First Three Minutes* (Deutsch, New York)

Weiskrantz L (1987) in *Mindwaves*, ed. Blakemore and Greenfield (Blackwell, Oxford) p.307

Weiskrantz L, Warrington E K, Sanders M D and Marshall J (1974) Brain **97** 709

Whitehead A N (1934) *Nature and Life* (CUP, Cambridge)

Whitehead A N and Russell B (1910) *Principia Mathematica*, 3 vols (CUP, Cambridge) (second edition, 1925)

Whitely C H (1962) Philosophy **37**, 61

Wigner E (1962) in *The Scientist Speculates*, ed. Good (Heinemann, London)

Williams D (1979) *Diffusions, Markov Processes, and Martingales* (Wiley, New York)

Index

9 780367 403270